# Mobile Computing

# Mobile Computing

Charles Harper

STATES
ACADEMIC PRESS
www.statesacademicpress.com

Published by States Academic Press,
109 South 5th Street,
Brooklyn, NY 11249, USA

ISBN: 978-1-63989-358-4

**Cataloging-in-Publication Data**

Mobile computing / Charles Harper.
    p. cm.
Includes bibliographical references and index.
ISBN 978-1-63989-358-4
1. Mobile computing. 2. Context-aware computing. 3. Portable computers.
4. Electronic data processing. I. Harper, Charles.
QA76.59 .M63 2022
004--dc23

For information on all States Academic Press publications
visit our website at www.statesacademicpress.com

# STATES
ACADEMIC PRESS

# TABLE OF CONTENTS

# PREFACE

This book is a culmination of my many years of practice in this field. I attribute the success of this book to my support group. I would like to thank my parents who have showered me with unconditional love and support and my peers and professors for their constant guidance.

The human-computer interaction, which allows transmission of data, video and voice is known as mobile computing. It uses computer or other wireless devices to transmit the data. Mobile computing includes mobile hardware and software along with mobile communication. Common forms of mobile computing devices are portable computers, smart cards, mobile phones and wearable computers. Mobile communication includes ad hoc networks, communication properties, data formats and infrastructure networks. Mobile application is the software application which is designed to operate mobile devices. Mobile hardware involves mobile device components that access or receive the service of mobility. Mobile computing has some fundamental principles that are portability, connectivity, individuality and social interactivity. This book aims to shed light on some of the unexplored aspects of mobile computing. Most of the topics introduced herein cover new techniques and the applications of mobile computing. This book will serve as a valuable source of reference for those interested in this field.

The details of chapters are provided below for a progressive learning:

Chapter – Introduction

Mobile computing deals with the transfer of data, voice and information over a network via computer or other wireless devices. Process migration, mobile agents, personal communication service, etc. are a few of its concepts. This chapter has been carefully written to provide an easy understanding of mobile computing.

Chapter – Cellular Communication

The communication technology enabling the use of mobile phones is referred to as cellular communication. Radio waves, microwaves and infrared waves are used in cellular communication. This chapter closely examines the concepts related to cellular communication to provide an extensive understanding of the subject.

Chapter – Wireless Networks

Wireless networks do not make use of wires and cables for establishing communication. It includes Bluetooth, hiperLAN and wireless PAN, LAN, MAN, WAN, and other wireless applications. The topics elaborated in this chapter will help in gaining a better perspective of the subject of wireless networks.

Chapter – Wireless Ad-Hoc Networks

Wireless ad-hoc network is a decentralized wireless network that does not depend on routers and access points. Multicasting, routing and mobile ad-hoc networks are studied under its domain. This chapter has been carefully written to provide an easy under-standing of wireless ad hoc networks.

Chapter – Issues and Security

Mobile communication network security is an essential component in mobile computing as it provides security of personal and business information stored in smartphones. Some of its aspects are Bluetooth security, wired equivalent privacy, Wi-Fi protected access, 4G and LTE network security. This chapter discusses the related aspects of security and issues of mobile computing in detail.

**Charles Harper**

# Introduction 1

- **Process Migration**
- **Mobile Agents**
- **Personal Communication Service**
- **Context Awareness**

Mobile computing deals with the transfer of data, voice and information over a network via computer or other wireless devices. Process migration, mobile agents, personal communication service, etc. are a few of its concepts. This chapter has been carefully written to provide an easy understanding of mobile computing.

Mobile computing is human–computer interaction by which a computer is expected to be transported during normal usage. Mobile computing involves mobile communication, mobile hardware, and mobile software. Mobile computing is the ability to use computing capability without a pre-defined location and connection to a network to publish and subscribe to information .Mobile computing as a generic term describing ability to use the technology to wirelessly connect to and use centrally located information and application software through the application of small, portable, and wireless computing and communication devices.

The term "Mobile computing" is used to describe the use of computing devices, which usually interact in some fashion with a central information system--while away from the normal, fixed workplace. Mobile computing technology enables the mobile worker to create, access, process, store communicate information without being constrained to a single location. By extending the reach of an organization's fixed information system, mobile computing enables interaction with organizational personnel that were previously disconnected. Mobile computing is the discipline for creating an information management platform, which is free from spatial and temporal constraints. The freedom from these constraints allows its users to access and process desired information from anywhere in the space.

The state of the user, static or mobile, does not affect the information management capability of the mobile platform being constrained to a single location.

## Different Types of Mobile Systems

The following section is an explanation of the different types of distributed systems ranging from the traditional type to nomadic, ad-hoc and finally ubiquitous ones.

### Traditional Distributed Systems

Traditional distributed systems consist of a collection of fixed hosts that are themselves attached to a network–if hosts are disconnected from the network this is considered to be abnormal whereas in a mobile system this is quite the norm. These hosts are fixed and are usually very powerful machines with fast processors and large amount of memory. Traditional distributed systems also need to guarantee non-functional requirements such as scalability, openness, heterogeneity, fault- and finally resource-sharing.

### Nomadic Distributed System

This kind of system is composed of a set of mobile devices and a core infrastructure with fixed and wired nodes. Mobile devices move from location to location, while maintaining a connection to the fixed network The mobile host has a home IP address and thus any packets sent to the mobile host will be delivered to the home network and not the foreign network where the mobile host is currently located. These systems are susceptible to the uncertainty of location, a repeated lack of connections and the migration into different physical and logical environments while operating.

### How Mobile Computing Work?

In mobile computing platform information between processing units flows through wireless channels. The processing units (client in client/server paradigm) are free from temporal and spatial constraints. That is, a processing unit (client) is free to move about in the space while being connected to the server. This temporal and spatial freedom provides a powerful facility allowing users to reach the data site (site where the desired data is stored) and the processing site (the geographical location where a processing must be performed) from anywhere. This capability allows organizations to set their offices at any location.

High-tier digital cellular systems include:

- Global System for Mobile Communications (GSM).
- IS -136 TDMA based Digital Advanced Mobile Phone Services (DAMPS).
- Personal Digital Cellular (PDC).
- IS -95 CDMA-based cdmaOne System.

Low-tier telecommunication systems include:

- Cordless Telephone 2 (CT2).

- Digital Enhanced Cordless Telephone (DECT).

- Personal Access Communication Systems (PACS).

- Personal Handy Phone Systems (PHS).

## Steps involved in Working of Mobile Computing

- The user enters or access data using the application on handheld computing device.

- Using one of several connecting technologies, the new data are transmitted from handheld to site's information system where files are updated and the new data are accessible to other system user.

- Now both systems (handheld and site's computer) have the same information and are in sync.

- The process work the same way starting from the other direction. The process is similar to the way a worker's desktop PC access the organization's applications, except that user's device is not physically connected to the organization's system. The communication between the user device and site's information systems uses different methods for transferring and synchronizing data, some involving the use of radio frequency (RF) technology.

## Characteristics of Mobile Computing

Mobile computing is accomplished using a combination of computer hardware, system and applications software and some form of communications medium. Powerful mobile solutions have recently become possible because of the availability of an extremely powerful and small computing devices, specialized software and improved telecommunication. Some of the characteristics of mobile computing are based on following:

## Hardware

The characteristics of mobile computing hardware are defined by the size and form factor, weight, microprocessor, primary storage, secondary storage, screen size and type, means of input, means of output, battery life, communications capabilities, expandability and durability of the device.

## Software

Mobile computers make use of a wide variety of system and application software. The most common system software and operating environments used on mobile computers includes MSDOS, Windows 3.1/3.11/95/98/NT, UNIX, android etc. These operating environments range in capabilities from a minimalist graphically-enhanced-pen-enabled DOS environment to the powerful capabilities of Windows NT. Each operating system/environment has some form of integrated development environment (IDE) for application development. Most of the operating environments provide more than one development environment option for custom application development.

## Communication

The ability of a mobile computer to communicate in some fashion with a fixed information system is a defining characteristic of mobile computing. The type and availability of communication medium significantly impacts the type of mobile computing application that can be created. The way a mobile computing device communicates with a fixed information system can be categorized as: (a) connected (b) weakly connected (c) batch and (d) disconnected. The connected category implies a continuously available high-speed connection. The ability to communicate continuously, but at slow 4peeds, allows mobile computers to be weakly connected to the fixed information system. A batch connection means that the mobile computer is not continuously available for communication with the fixed information system. Mobile computers may operate in batch mode over communication mediums that are capable of continuous operation, reducing the wireless airtime and associated fees. Disconnected mobile computers allow users to improve efficiency by making calculations, storing contact information, keeping a schedule, and other non-communications oriented tasks.

## Applications and Benefits of Mobile Computing

The real power of mobile computing becomes apparent when mobile hardware, Software, and communications are optimally configured and used to accomplish a Specified mobile task. Although many varied applications exist, mobile computing applications can generally be divided into two categories--horizontal and vertical.

## Horizontal

Horizontal applications have broad-based appeal and include software that performs functions such as: (a) email; (b) Web browsing; (c) word processing; (d) scheduling;

(e) contact management; (f) to-do lists; (g) messaging; (h) presentation. These types of applications usually come standard on Palmtops, Clamshells, and laptops with systems software such as Windows 95.

## Vertical

Vertical applications are industry-specific and only have appeal within the specific Industry for which the application was written. Vertical applications are commonly used in industries such as: (a) retailing; (b) utilities; (c) warehousing; (d) shipping; (e) Medical and (f) law enforcement and public safety. These vertical applications are often transaction oriented and normally interface with a corporate database. Other application areas include: (a) mining; (b) forestry; (c) agriculture; and (d) surveying etc.

## Benefits of Mobile Computing

Mobile computing technology offers a quick and easy way to increase efficiency, productivity and profitability while gaining better control of our operations. The power and data storage capacity of today's handheld PCs and Personal Digital Assistants (PDAs) has made low-cost mobile computing a practical reality. Today's world mobile computing is using in various fields.

## Improved Information Accessibility

Mobile computing enables improvements in information accessibility. The degree of improvement is directly dependent upon the mobile hardware and communications equipment in use. Mobile computing technology (hardware, software, and communications) provides a wide range of options that can be mixed and matched to fit the needs of each individual mobile computing application. The improvements in information accessibility enabled by mobile computing result in improved information flow both to and from the central fixed information system. The mobile computer enables quick and efficient information retrieval from the central information system. The ability to access central information and make fixed or ad hoc queries of corporate databases enables employees to get the information they need to complete the job. The mobile computer also enables transmission of current operational data, in native digital format, from the mobile user to the central fixed information system. Once transmitted to the fixed information system, the data from the mobile user can be processed and made available for all other users of the central information system. Thus, the information available to a mobile user from the central information system reflects current information from other mobile users as well. Mobile computing eliminates the delay that occurs when an employee must physically return to the office at the end of the day and submit paper forms so that data entry personnel can enter the information into the central information system. Even employees that are not continuously connected to the fixed organizational information system via a wireless link will experience significantly improved information accessibility through mobile computing. One phone call at the end of the day from the mobile user via a standard modem is all that is required to transmit the entire day's transactions to the central computer, saving travel and data entry time. Additionally, any scheduling or assignment changes for the mobile employee for the following day can be transmitted to the employee during the same phone call. Mobile computing also significantly speeds information accessibility when other media, such as: (a) facsimile; (b) audio files; or (c) still images are concerned. Digital images or audio files can be accessed by the mobile user or transmitted from the mobile user to the central fixed organizational information system. If matched properly to the work environment and task to be accomplished, the mobile computer will always be in the possession of the mobile worker during the course of the day. Especially in the connected or weakly connected modes of operation, this means that the mobile employee may be contacted throughout the workday via the mobile computing device. Additionally, it means that the employee has access to other mobile employees via email or other messaging schemes. As with many mobile computing applications, the type of mobile application and the hardware, software, and communications used to support it will normally determine the degree and type of information accessibility. The direct measurable results of improved information accessibility-both to and from the mobile worker are many. They include: (a) improved customer service; (b) reduced cycle times; (c) greater accuracy; (d) fewer complaints; and (e) a reduction in required intermediate support staff.

## Increased Operational Efficiency

Mobile computing enables improvements in the operational efficiency of organizations that integrate the technology into their fixed information systems. It enables the computing power and information contained within the fixed information system to be structured around the optimum work flow of a mobile worker, instead of altering the mobile worker's work flow to meet the optimum configuration for computing. The mobile computer stays with the mobile employee, instead of the employee being required to travel to the computer. Mobile computing can improve efficiency in many ways, including: (a) saving time; (b) reducing waste; (c) cutting cycle times; (d) reducing rework; (e) enabling business process reengineering; (f) improving accuracy; (g) decreasing time spent on customer complaints; and (h) reducing unnecessary travel.

## Increased Management Effectiveness

Mobile computing technology can improve management effectiveness by improving information quality, information flow, and ability to control a mobile workforce. It makes the most current and accurate information available to both the mobile worker and the users of the fixed information system with which the mobile worker communicates. These benefits can be seen in all areas of the information System, often, it is the improved ability to manage operations that is partly responsible for the performance improvements seen in companies that introduce mobile computing technology.

## For Estate Agents

Estate agents can work either at home or out in the field. With mobile computers they can be more productive. They can obtain current real estate information by accessing multiple listing services, which they can do from home, office or car when out with clients. They can provide clients with immediate feedback regarding specific homes or neighborhoods, and with faster loan approvals, since applications can be submitted on the spot. Therefore, mobile computers allow them to devote more time to clients.

## Emergency Services

Ability to receive information on the move is vital where the emergency services are involved. Information regarding the address, type and other details of an incident can be dispatched quickly, via a Cellular Digital Packet Data (CDPD) system using mobile computers, to one or several appropriate mobile units, which are in the vicinity of the incident.

## In courts

Defense counsels can take mobile computers in court. When the opposing counsel references a case which they are not familiar, they can use the computer to get direct, real-time access to on-line legal database services, where they can gather information on the case and related precedents. Therefore mobile computers allow immediate access to a wealth of information, making people better informed and prepared.

## In Companies

Managers can use mobile computers in, say, and critical presentations to major customers. They can access the latest market share information. At a small recess, they can revise the presentation to take advantage of this information. They can communicate with the office about possible new offers and call meetings for discussing responds to the new proposals. Therefore, mobile computers can leverage competitive advantages.

## Credit Card Verification

At Point of Sale (POS) terminals in shops and supermarkets, when customers use credit cards for transactions, the intercommunication is required between the bank central computer and the POS terminal, in order to effect verification of the card usage, can take place quickly and securely over cellular channels using a mobile computer unit. This can speed up the transaction process and relieve congestion at the POS terminals.

## Field Sales

The operational efficiency of sales personnel is significantly enhanced through mobile computing. An excellent example of these improvements can be seen by examining how mobile computing improves the efficiency of remote insurance and financial planning sales. The mobile computer frees the sales agent to meet with the client at the client's home, office, or other location. Customer data is collected, estimates and comparisons are immediately calculated, the customer decides on the program of choice, the central computer is immediately updated, and the customer is enrolledin the insurance or financial planning program. Without mobile computing, this sales process would take days instead of minutes. In addition to accessing and updating customer account information, mobile sales personnel can accomplish tasks such as printing invoices or other information to leave with the customer.

## Transportation and Shipping

Using mobile computers in conjunction with GPS/GIS and an accompanying vehicle information system (VIS), the operations of an entire transportation fleet can be managed from a central location. The central office knows the location, status, and condition of all vehicles, and operators have two-way communication with the operations center. Using this information, vehicles can be optimally dispatched to maximize efficiency as measured by: (a) time; (b) fuel consumption; and (c) delivery priority. The mobile computers enable significant performance improvements, achieved simultaneously with operational cost reductions.

## General Dispatching

Mobile computers used in conjunction with Global Positioning System (GPS) and Geographical Information System (GIS) data allow significant improvements in the operational efficiency of various dispatch operations. For example, the central computer at a taxi company can track the location and status of all its taxicabs and electronically dispatch the most appropriate car to a customer's location. Additionally, the central computer can calculate an accurate approximation of when the taxi will arrive, enabling improved customer service.

## Hotel Operations

Connecting the cleaning and hospitality staff of a hotel with mobile computing can significantly improve the efficiency of hotel operations. As guests check out and rooms are vacated, the central computer wirelessly signals cleaning staff that the rooms are ready for cleaning. Problems that are identified during cleaning, such as broken appliances or faulty plumbing, are immediately communicated to the mobile maintenance team for action. As soon as cleaning is complete and repairs are accomplished, the cleaning staff member wirelessly updates the central computer and the room is immediately available for check-in by a new guest. The same system can be used to efficiently direct mobile hospitality personnel in response to guest requests for information and service.

## News Reporting

Mobile computers dramatically improve the efficiency of news media operations. Reporters equipped with mobile computers and accompanying electronic devices can cover news or sporting event, take digital video or still photographs, digitally record audio interviews, compose the text of the news story, and transmit the completed product back to the central agency for editing and immediate publication. In the media industry, the timing and quality of news coverage is critical. Mobile computing increases the quality of the information from the media crews and significantly decreases the time required to process and transmits the story for publication.

## Health Care

Mobile medical care, whether in-home, on the road, or within a hospital, is more efficient with mobile computing. The mobile healthcare worker can access patient records for reference purposes, and is able to update records with current diagnosis and treatment information. Emergency medical technicians (EMTs) responding at the scene of an accident can use mobile computers to capture patient information, treatments accomplished vital signs, and other critical data. This information is wirelessly transmitted to the receiving hospital, which then prepares to receive and treat the patient, or recommend another hospital facility with more appropriate treatment facilities depending upon the nature and severity of the injuries. The more efficient hand-off between ambulance EMTs and hospital staff made possible by mobile computing can save lives that otherwise might have been lost.

## Fieldwork

Almost any form of fieldwork can be made significantly more efficient through the application of mobile computing. Parking control officers and utility inspectors are two examples of field workers who can receive operational benefits from mobile computing. Parking control officers use handheld computers to check the registration and violation history of parking offenders. Parking violations are issued immediately and towing/backup can be requested when required. Utility inspectors have historically used paper forms to capture information such as consumer power consumption and utility equipment status (transformers, transmission towers, etc.). Using mobile computers, inspectors can be given instructions on inspections to be accomplished and information can be captured and validated at the source.

## Mobile Automation

General business travelers also reap the benefits of mobile computing. E-mail, Spreadsheets, presentations, and word processing are the four primary tasks accomplished by these business travelers. Laptops, Palmtops, and portable Clamshell computers with usable-size keyboards enable businesspeople to stay in touch and accomplish the tasks they need for job effectiveness. Using powerful mobile computers in conjunction with high-speed connectivity, mobile workers can perform work normally accomplished at the office while on the road or in the field.

Just as mobile computing enables improved operational efficiency, it also enables improved management effectiveness. Mobile computers make more timely and accurate information available to managers. Mobile computers improve the manager's ability to track work in progress. They also improve the ability of managers to communicate with mobile personnel. Additionally, mobile computers provide better information to mobile employees, so they can make more informed decisions locally and minimize the need for management decisions from the central office.

## Process Migration

Process Migration refers to the mobility of executing (or suspended) processes in a distributed computing environment. Usually, this term indicates that a process uses a network to migrate to another machine to continue its execution there. Sometimes the term is used to describe the change in execution from one processor to another processor within the same machine. I will use process migration in the first context programs migrating between machines.

Several problems occur when a running process moves to another machine. Some of these problems are:

I/O redirection: if a process does I/O to files or devices that are bound to a certain machine, there must be a way to redirect access to these resources even after the process migrated. This involves redirection of the I/O data stream over the network and has disadvantages concerning security, performance and reliability.

Inter-process communication: messages sent to a process with process ID **P** on a machine **M** have to be redirected to the new machine **N** and the new process ID **Q**. The machine the process migrated away from needs to keep records of migrated processes. If multiple migration occurs, the overhead increases.

Shared memory: if one of a group of cooperating processes migrates away and all these processes use a shared memory segment, then the network must be used to emulate shared memory access. This adds complexity and slows down access to the shared memory dramatically for processes that migrated away from the machine holding the shared memory.

The phenomena that a host computer must provide services to a process that migrated away is called *Residual Dependency*.

## Technology behind Process Migration

In general, the following things are required to allow agents to migrate across a network:

- Common execution language.

- Process persistence.

- Communication mechanism between agent hosts.

- Security to protect agents and agent hosts.

## Common Execution Language

If a process is to migrate from one host to another, then both hosts must share a common execution language. In a homogenous networking environment, it's conceivable that assembly language or machine code could be sent across the network for execution. However, such a system would be extremely limited, and not very future proof.

A more likely scenario for mobile agency is a heterogeneous environment, where many different system architectures are connected. In this case, an interpreted scripting language or emulation of a system that is capable of executing machine code  solves the problem of a common execution language.

## Process Persistence

For processes to migrate to remote machines, they must be capable of saving their execution state, or spawning a new process whose execution state will be saved. This property is called persistence. Persistence involves converting the object's state (variables, stack, and possibly even the point of execution) and converting it into a data form suitable for transmission over a network. Agents should not have to be responsible for achieving this themselves, and process persistence would likely be built into the mobile agent language or architecture.

## Communication Mechanism between Agent Hosts

Some communication mechanism must exist to transfer agents across networks. An agent might be transferred using TCP/IP, or by using a higher level of communication such as RMI, IIOP, SMTP or even HTTP. Mobile agent architectures may even use a variety of transport mechanisms, giving greater flexibility.

An agent's executable code must be transferred, which may consume a large amount of network bandwidth, unless shared code is located at the agent host. Techniques such as shared libraries of code, or caching, may be of benefit. In addition, the persistent state of the agent must be transferred.

## Security to Protect Agents and Agent Hosts

Security is critical when executable code is transferred across a network. Malicious or badly written code could wreak havoc when unleashed upon an unsuspecting host, and agents themselves need protection against hostile hosts that would seek to dissect or modify them. There is no magic solution that will solve all the security problems of mobile agents, but precautions can be taken to minimize risk.

When an agent leaves for a new host, extreme care must be taken to prevent unauthorized modification or analysis of the agent. Agents may carry with them confidential or sensitive information and logic, which shouldn't be accessible to the agent host. Encryption may be of benefit, but the data and code must be decrypted at some point in time for the agent to execute. Once this occurs, the agent becomes vulnerable, and is at the mercy of the agent host. In a scripting language, the internal logic of the agent is exposed, but even compiled languages can be decompiled with a disturbing degree of success .Other than using trusted hosts, there is little that can be done to protect the agent from snooping eyes.

For agent hosts, the news is a little more positive. Through the use of digital signatures, the identity of an agent and its user can be authenticated. However, in many commercial environments, even un-trusted agents will need to be accepted. These agents may carry with them dangerous payloads, such as code that needlessly consumes CPU time and memory, that attempts to damage the agent host or carries back information that would make it vulnerable to hacking. Such is the risk an agent host must take, though limiting access to resources such as disks and the network may offer some solution.

## Implications of Mobile Agents

If mobile agents were to gain widespread commercial adoption (to the degree that say, web browsing or email has), then what type of a network would we have? The following list of implications is by no means exhaustive, but does provide an interesting set of observations.

## Bandwidth Conservation

One of the goals of mobile agency is to conserve bandwidth, by placing an agent directly at the point of information, rather than sending dozens or even hundreds of queries across the network. This is based on the premise that these queries would have consumed more bandwidth than sending an agent over the network, and bringing it back again .

Bandwidth conservation is an admirable goal, but whether mobile agents will help realize this goal is questionable. For bandwidth to be conserved, the bandwidth consumed by sending across a mobile agent, and waiting for its results, must be less than that of a series of queries sent via a messaging or RPC system. This is a determination that must be made in practice, and cannot be fully verified just by theory.

Many scenarios can be foreseen. Perhaps mobile agents could comb through large amounts of resources on a single site, and bring back a small number of matches, in a similar nature to a search engine. However, some electronic commerce models suggest that mobile agents would be sent out to multiple sites, perhaps to negotiate low prices with vendors (a shopping-bot). This sort of activity has the potential to result in an incredible explosion of bandwidth consumption.

Indeed, searching is a task that many people would like to see mobile agents performing. Currently, a small number of indexing agents collect information for search engines, while millions of queries are made by users. Imagine if the same number of queries were made instead by mobile agents that traveled across the network to sites. Two scenarios are possible. Either a much larger amount of bandwidth (and CPU usage for the agent hosts) will be consumed, or a much lesser amount of bandwidth will be consumed as users receive more accurate search results because their agents have more control over the search process. Instinct suggests, however, that a simple keyword query entered via a web browser will consume less resources and bandwidth than sending an agent with specialised searching algorithms across the network.

## Delegate Tasks to Agents when not Connected

The Internet, as it stands today , is made up of many millions of computers, some of which are permanently connected but the majority of which connect via dial-up modem connections for short periods of times. Imagine if you could delegate tasks to mobile agents, that would roam the network for you while not connected. This goal would be extremely desirable in the short term, until permanent connections became more prevalent. Here mobile agents may have found a sound market. This market could potentially be profitable, for the mobile agent technology vendors and the agent hosts that allow offline usage of their network.

Delegation of tasks to mobile agents could also be used as a form of load sharing in distributed systems. Agents could perform tasks on remote systems, moving from system to system as required to balance the load. Mobile agency also gives greater flexibility, because new tasks and new code can be added to the system without the need for a fixed codebase.

## Mobile Agents Enable New Types of Interaction

The ability of mobile agents to fragment themselves into many pieces that travel to different points across the network sounds promising. It might enable new forms of interaction, such as negotiating agents that travel to vendors seeking the best deal, or meeting places where agents can "get together" and communicate. The attraction of mobile agents for electronic commerce is great, and it might make sense to deploy mobile agents for electronic commerce. However, such uses could also be accomplished by message passing, or direct communication using application protocols like HTTP. Mobile agency is promising, but it is not the only mechanism for new uses of software agents.

## Agent Privacy

If mobile agents were to become commonplace, serious privacy concerns would be raised. Aside from deliberate attempts to decompile or interrogate an agent (or its encrypted data), agent hosts could also monitor the actions of agents, and create consumer profiles. Even knowledge of the types of queries, or the way in which an agent searched, could reveal information about its owner. When individuals query a search engine, there is some degree of anonymity, but there is less control with mobile agency .

## Competing Technologies

Mobile agency faces stiff competition from other technologies that can achieve similar outcomes, such as message passing or advanced forms of remote procedure calls. Some of the more notable technologies that offer competition to mobile agency are discussed.

### Message Passing Systems

Software agents need not always travel across a network to communicate with information sources, or other agents. One of the most important message passing systems for agents is the Knowledge Query Manipulation Language (KQML). KQML is an important mechanism for communication, because it allows much more complex forms of interaction than query/response mechanisms. KQML allows agents to communicate using a rich set of messages called performatives, and is capable of communicating attitudes about information, rather than just data and facts. Message passing systems like KQML don't require mobility, they can simply pass a message and have it delivered through some transport mechanism.

### Remote Method Invocation

Remote method invocation allows Java developers to write distributed systems that share objects. New objects can be transferred across the network, and RMI is becoming a popular mechanism for agent communication. RMI can be used to facilitate mobile agency (acting as a transport mechanism), or as a replacement that allows agents to invoke methods of other agents.

### Common Object Request Broker Architecture

CORBA is a platform and language independent mechanism for invoking remote object methods. Unlike RMI which is specific only to Java Virtual Machines, CORBA can be used to create distributed systems that execute on many platforms, in many languages.

CORBA holds great potential, because of its portability and flexibility. CORBA is a direct threat to mobile agency, and would allow developers to create agents that are capable of complex communication without ever traveling across a network.

## Mobile Agents

More specifically, a mobile agent is a process that can transport its state from one environment to another, with its data intact, and be capable of performing appropriately in the new environment. Mobile agents decide when and where to move. Movement is often evolved from RPC methods. Just as a user directs an Internet browser to "visit" a website (the browser merely downloads a copy of the site or one version of it in the case of dynamic web sites), similarly, a mobile agent accomplishes a move through data duplication. When a mobile agent decides to move, it saves its own state, transports this saved state to the new host, and resumes execution from the saved state. A mobile agent is a specific form of mobile code.

However, in contrast to the Remote evaluation and Code on demand programming paradigms, mobile agents are active in that they can choose to migrate between computers at any time during their execution. This makes them a powerful tool for implementing distributed applications in a computer network. An open multi-agent system (MAS) is a system in which agents that are owned by a variety of stakeholders continuously enter and leave the system. Mobile agents decide when and where to move.

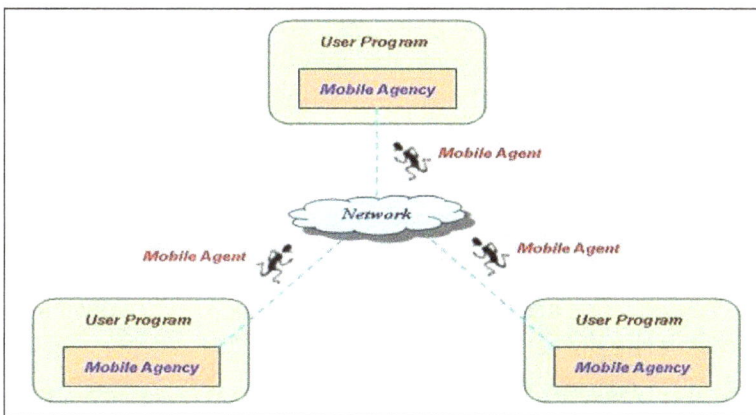

Mobile agent.

The appeal of mobile agents is quite alluring mobile agents roaming the Internet could search for information, find us great deals on goods and services, and interact with other agents that also roam networks (and meet in a gathering place) or remain bound to a particular machine. Significant research and development into mobile agency has been conducted in recent years, and there are many mobile agent architectures available today. During migrating to the unreliable network link, an agent can continue executing if the network link goes down with the help like Genetic algorithm and neural networks, making mobile agents particularly attractive in mobilecomputing environments. The true strength of mobile agents is not that they make new distributed applications possible, but rather that they allow a wide range of distributed applications to be implemented robustly and easily within a single, general framework.

## Working of Mobile Agents

A mobile agent consists of the program code and the program execution state (the current values of variables, next instruction to be executed, etc.). Initially a mobile agent resides on a computer called the home machine. The agent is then dispatched to execute on a remote computer called a mobile agent host (a mobile agent host is also called mobile agent platform or mobile agent server). When a mobile agent is dispatched the entire code of the mobile agent and the execution state of the mobile agent is transferred to the host. The host provides a suitable execution environment for the mobile agent to execute. The mobile agent uses resources (CPU, memory, etc.) of the host to perform its task. After completing its task on the host, the mobile agent migrates to another computer. Since the state information is also transferred to the host, mobile agents can resume the execution of the code from where they left off in the previous host instead of having to restart execution from the beginning. This continues until the mobile agent returns to its home machine after completing execution on the last machine in its itinerary.

## Properties of Mobile Agents

Mobile agents have the following unique properties:

- Adaptive Learning: Mobile agents can learn from experiences and adapt themselves to the environment. They can monitor traffic in large networks and learn about the trouble spots in the network. Based on the experiences of the agent in the network the agent can choose better routes to reach the next host.

- Autonomy: Mobile agents can take some decisions on its own. For example, mobile agents are free to choose the next host and when to migrate to the next host. These decisions are transparent to the user and the decisions are taken in the interest of the user.

- Mobility: Mobile agents have the ability to move from one host to another in the network.

## Advantages of Mobile Agents

- Reduction in Network Load: The interactions in a distributed system are often achieved using communication protocols. These protocols involve transfer of large volumes of data stored at remote hosts over the network to a central processing site resulting in high network traffic. An alternative to using communication protocols is the use of mobile agents. Mobile agents are dispatched to the remote hosts containing the data. The agents perform the computations at the remote hosts and return back with the results. Since computations are moved to the data storage location instead of moving data to the computing location, network load is reduced.

- Overcome Network Latency: Consider a manufacturing plant in which many critical real time systems are controlled through a network. Controlling many systems through a network involves significant delays, which are not acceptable for critical real time systems. To overcome this problem, mobile agents can be directly dispatched from the central controller in the manufacturing plant to the real time systems. The agents act locally and directly execute the controller's directions.

- Protocol Encapsulation: Protocols enable components of a distributed system to communicate and co-ordinate their activities. However, protocols evolve over a period of time and new features such as better security may be introduced in the protocol. It is a cumbersome task to upgrade the protocol code at all locations in the distributed system. Mobile agents offer a solution to this problem. The mobile agent code can encapsulate the protocol. When a protocol is upgraded, only the mobile agent has to be altered.

- Asynchronous and Autonomous Execution: Mobile agents operate asynchronously. Once a mobile agent is dispatched from the home machine, the home machine can disconnect from the network. The mobile agent executes autonomously without the intervention of the home machine. The home machine can reconnect at a later time and collect the agent.

- Fault Tolerance: Mobile agents react dynamically and autonomously to the changes in their environment, which makes them robust, and fault tolerant. They have the ability to distribute themselves in the network in such a way as to maintain the optimal configuration for solving the particular problem. If a host is being shutdown, all agents executing on that machine will be warned and given time to dispatch themselves and continue their operation on another host in the network.

## Disadvantages of Mobile Agents

The main drawback of mobile agents is the security risk involved in using mobile agents. Security risks in a mobile computing environment are twofold. Firstly a malicious mobile agent can damage a host. For example a virus can be disguised as a mobile agent and distributed in the network causing damage to the host machines that execute the agent. On the other hand a malicious host can tamper with the functioning of the mobile agent. Most experts suggest that this risk is far more difficult to deal with. To illustrate this scenario, consider a mobile agent that visits the servers of several airlines to buy a ticket for the lowest price. A malicious airline server can try to obtain sensitive price information from the mobile agent. The malicious server may tamper with the mobile agent and increase the prices quoted by other airlines thereby giving it an unfair advantage. Some servers may even try to steal the credit card number from the mobile agent. The defense mechanisms suggested try to prevent malicious actions in the first place. These include safe programming languages that prevent mobile agents from performing malicious actions on the hosts. If a malicious action does occur then the defense mechanisms detect it as soon as possible and take remedial action. One of the schemes proposed is to introduce a tracing mechanism that records the execution of the mobile agent at each host. When the agent is dispatched to the next host, the trace is also sent. Using this trace, malicious actions can be detected and the malicious host can be identified. However, in spite of significant development in the field of cryptography there still exist many security issues that need to be addressed in mobile agents.

## Applications of Mobile Agents

Although no universally used application (normally called killer application) has been developed for them, mobile agents are suitable for the following applications:

- Parallel Computing: Solving a complex problem on a single computer takes a lot of time. To overcome this, mobile agents can be written to solve the problem. These agents migrate to computers on the network, which have the required resources and use them to solve the problem in parallel thereby reducing the time required to solve the problem.

- Data Collection: Consider a case wherein, data from many clients has to be processed. In the traditional client-server model, all the clients have to send their data to the server for processing resulting in high network traffic. Instead mobile agents can be sent to the individual clients to process data and send back results to the server, thereby reducing the network load.

- E-commerce: Mobile agents can travel to different trading sites and help to locate the most appropriate deal, negotiate the deal and even finalize business transactions on behalf of their owners. A mobile agent can be programmed to bid in an online auction on behalf of the user. The user himself need not be online during the auction.

- Mobile Computing: Wireless Internet access is likely to stay slow and expensive. Power consumption of wireless devices and high connection fee deter users from staying online while some complicated query is handled on behalf of the user. Users can dispatch a mobile agent, which embodies their queries, and log off, and the results can be picked up at a later time.

## Information Gathering and Retrieval by Mobile Agent

Mobile agents introduce new levels of complexity, operating within an environment that is autonomous, open to security attacks, agent server crashes, and failure to locate resources. There is significant attention within the mobile agent fault tolerance community concerning the loss of mobile agents at remote agent servers that fail by crashing. If a user may dispatch a mobile agent to visit airlines to find the cheapest flight. The mobile agent migrates to each airline's remote agent server and dynamically updates its own internal state to reflect the cheapest price, unfortunately if agent server crashes, all information relating to the cheapest flight is lost. Some applications introduce failure dependencies with agent servers. Consequently transaction-based solutions introduce unnecessary performance overheads.

Some techniques modify the agent server platform to introduce fault tolerance, e.g. an agent server may replicate each mobile agent before execution a mobile agent inject a replica into stable storage upon arriving at an agent server. However, in the event of an agent server crash, the replica remains unavailable for an unknown time period. Introducing fault tolerance into the agent server platform restricts information retrieval to those enterprises that allow the modified agent server platform.

## Genetic Algorithm

Genetic Algorithms (GAs) are adaptive heuristic search algorithm premised on the evolutionary ideas of natural selection and genetic. The basic concept of GAs is designed to simulate processes in natural system necessary for evolution, specifically those that follow the principles first laid down by Charles Darwin of survival of the fittest. As such they represent an intelligent exploitation of a random search within a defined search space to solve a problem.

Not only does GAs provide alternative methods to solving problem, it consistently outperforms other traditional methods in most of the problems link. Many of the real world problems involved finding optimal parameters, which might prove difficult for traditional methods but ideal for GAs. However, because of its outstanding performance in optimization, GAs has been wrongly regarded as a function optimizer. In fact, there are many ways to view genetic algorithms. Perhaps most users come to GAs looking for a problem solver, but this is a restrictive view. Herein, we will examine GAs as a number of different things:

- GAs as problem solvers.

- GAs as challenging technical puzzle.

- GAs as basis for competent machine learning.

- GAs as computational model of innovation and creativity.

- GAs as computational model of other innovating systems.

- GAs as guiding philosophy.

Genetic algorithms are one of the best ways to create a high quality solution. Genetic algorithms use the principles of selection and evolution to produce several solutions to a given problem. When traditional search methods fail, we employ genetic algorithm (GA) to search for the near-optimal solution. Genetic algorithm helps in handoff procedure while node is changing their cluster in VANET. An effective GA representation and meaningful fitness evaluation are the keys of the success in GA applications. The appeal of GAs comes from their simplicity and elegance as robust search algorithms as well as from their power to discover good solutions rapidly for difficult high-dimensional problems. GAs is useful and efficient for large search space and no mathematical analysis is available.

A genetic algorithm is an adaptive heuristic search program that applies the principles of evolution found in nature. Genetic algorithm combines selection, crossover, and mutation operators with the goal of finding the solution of best fitness to a problem. A genetic algorithm searches for the optimal solution until a specified termination criterion is met. The solution to a problem is called a chromosome. A chromosome is made up of a collection of genes which are simply the parameters to be optimized.

## Benefit from Genetic Algorithm

An effective GA representation and meaningful fitness evaluation are the keys of the success in GA applications. The appeal of GAs comes from their simplicity and elegance as robust search algorithms as well as from their power to discover good solutions rapidly for difficult high-dimensional problems. GAs is useful and efficient when:

- The search space is large, complex or poorly understood.
- Domain knowledge is scarce or expert knowledge is difficult to encode to narrow the search space.
- No mathematical analysis is available.
- Traditional search methods fail.

The advantage of the GA approach is the ease with which it can handle arbitrary kinds of constraints and objectives; all such things can be handled as weighted components of the fitness function, making it easy to adapt the GA scheduler to the particular requirements of a very wide range of possible overall objectives.

## Genetic Algorithm Operations

The genetic algorithm performs the following operations:

Initialization: Initially many individual solutions are randomly generated to form an initial population. The population size depends on the nature of the problem. Traditionally, the population is generated randomly, covering the entire range of possible solutions.

Selection: During each successive generation, a proportion of the existing population is selected to breed a new generation. Individual solutions are selected through a fitness-based process, where fitter solutions are typically more likely to be selected.

Reproduction: The next step is to generate a second generation population of solutions from those selected through genetic operators: crossover (also called recombination), and/or mutation. For each new solution to be produced, a pair of "parent" solutions is selected for breeding from the pool selected previously. By producing a "child" solution using the above methods of crossover and mutation, a new solution is created which typically shares many of the characteristics of its "parents". New parents are selected for each new child, and the process continues until a new population of solutions of appropriate size is generated. Although Crossover and Mutation are known as the main genetic operators, it is possible to use other operators such as regrouping, colonization-extinction, or migration in genetic algorithms.

Termination: This generational process is repeated until a termination condition has been reached. Common terminating conditions are:

- A solution is found that satisfies minimum criteria.

- Fixed number of generations reached.

- Allocated budget (computation time/money) reached.

- The highest ranking solution's fitness is reaching or has reached a plateau such that successive iterations no longer produce better results.

- Manual inspection.

- Combinations of the above.

## Personal Communication Service

PCS (personal communications service) is a wireless phone service similar to cellular telephone service but emphasizing personal service and extended mobility. It's sometimes referred to as *digital cellular* (although cellular systems can also be digital). Like cellular, PCS is for mobile users and requires a number of antennas to blanket an area of coverage. As a user moves around, the user's phone signal is picked up by the nearest antenna and then forwarded to a base station that connects to the wired network. The phone itself is slightly smaller than a cellular phone. According to Sprint, PCS is now available to 230 million people.

The "personal" in PCS distinguishes this service from cellular by emphasizing that, unlike cellular, which was designed for car phone use and coverage of highways and roads, PCS is designed for greater user mobility. It generally requires more cell transmitters for coverage, but has the advantage of fewer blind spots. Technically, cellular systems in the United States operate in the 824-849 megahertz (MHz) frequency bands; PCS operates in the1850-1990 MHz bands.

Several technologies are used for PCS in the United States, including Time Division Multiple Access (TDMA), Code Division Multiple Access (CDMA), and Global System for Mobile (GSM) communication. GSM is more commonly used in Europe and elsewhere.

## Context Awareness

Context awareness is the ability of a system or system component to gather information about its environment at any given time and adapt behaviors accordingly. Contextual or context-aware computing uses software and hardware to automatically collect and analyze data to guide responses.

Context includes any information that's relevant to a given entity, such as a person, a device or an application. As such, contextual information falls into a wide range of categories including time, location, device, identity, user, role, privilege level, activity, task, process and nearby devices/users.

Web browsers, cameras, microphones and Global Positioning Satellite (GPS) receivers and sensors are all potential sources of data for context-aware computing. A context-aware system may gather data through these and other sources and respond according to pre-established rules or through computational intelligence. Such a system may also base responses on assumptions about context. For user applications, context awareness can guide services and enable enhanced experiences including augmented reality, context-relevant information delivery and contextual marketing messages.

Although often defined as a property of mobile devices used to present relevant, actionable information to the end user, context awareness is also a technological driver for M2M (machine to machine) and Internet of Things (IoT), ubiquitous computing and event-driven computing environments.

### References

- Applications-and-Benefits-of-Mobile-Computing: ijser.org, Retrieved 28 April, 2020
- Mobile-agents, software-agents: davidreilly.com, Retrieved 05 January, 2020
- PCS: searchnetworking.techtarget.com, Retrieved 23 July, 2020
- Context-awareness: whatis.techtarget.com, Retrieved 14 June, 2020

- **Electromagnetic Spectrum**

- **Radio Waves**

- **Microwaves**

- **Infrared Waves**

- **Communication Satellites**

- **First Generation (1G)**

- **Second Generation (2G)**

- **Third Generation (3G)**

- **Fourth Generation (4G)**

- **Fifth Generation (5G)**

The communication technology enabling the use of mobile phones is referred to as cellular communication. Radio waves, microwaves and infrared waves are used in cellular communication. This chapter closely examines the concepts related to cellular communication to provide an extensive understanding of the subject.

All cellular telephone systems exhibit several fundamental characteristics, as summarized in the following:

- The geographic area served by a cellular system is broken up into smaller geographic areas, or cells. Uniform hexagons most frequently are employed to represent these cells on maps and diagrams; in practice, though, radio waves do not confine themselves to hexagonal areas, so the actual cells have irregular shapes.

- All communication with a mobile or portable instrument within a given cell is made to a base station that serves the cell.

- Because of the low transmitting power of battery-operated portable instruments, specific sending and receiving frequencies assigned to a cell may be reused in other cells within the larger geographic area. Thus, the spectral efficiency of a cellular system (that is, the uses to which it can put its portion of the radio spectrum) is increased by a factor equal to the number of times a frequency may be reused within its service area.

- As a mobile instrument proceeds from one cell to another during the course of a call, a central controller automatically reroutes the call from the old cell to the new cell without a noticeable interruption in the signal reception. This process is known as handoff. The central controller, or mobile telephone switching office (MTSO), thus acts as an intelligent central office switch that keeps track of the movement of the mobile subscriber.

- As demand for the radio channels within a given cell increases beyond the capacity of that cell (as measured by the number of calls that may be supported simultaneously), the overloaded cell is "split" into smaller cells, each with its own base station and central controller. The radio-frequency allocations of the original cellular system are then rearranged to account for the greater number of smaller cells.

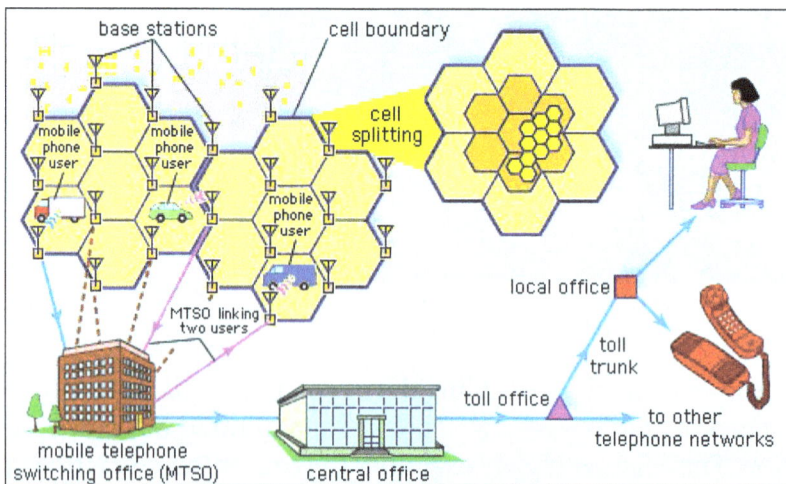

Operation of a cellular telephone system.

From a specific location within a geographic area, or cell, a subscriber places a call using a mobile telephone. The call is relayed by the base station serving that cell to the mobile telephone switching office (MTSO). The MTSO in turn relays the call to another base station within the cellular system or to a central office in the public switched telephone network. When telephone traffic within a cell exceeds capacity, the cell is split into a number of smaller cells, each with its own base station.

Frequency reuse between discontiguous cells and the splitting of cells as demand increases are the concepts that distinguish cellular systems from other wireless telephone systems. They allow cellular providers to serve large metropolitan areas that may contain hundreds of thousands of customers.

## Development of Cellular Systems

In the United States, interconnection of mobile transmitters and receivers with the public switched telephone network (PSTN) began in 1946, with the introduction of mobile telephone service (MTS) by the American Telephone & Telegraph Company (AT&T). In the U.S. MTS system, a user who wished to place a call from a mobile phone had to search manually for an unused channel before placing the call. The user then spoke with a mobile operator, who actually dialed the call over the PSTN. The radio connection was simplex i.e., only one party could speak at a time, the call direction being controlled by a push-to-talk switch in the mobile handset. In 1964 AT&T introduced the improved mobile telephone service (IMTS). This provided full duplex operation, automatic dialing, and automatic channel searching. Initially 11 channels were provided, but in 1969 an additional 12 channels were made available. Since only 11 (or 12) channels were available for all users of the system within a given geographic area (such as the metropolitan area around a large city), the IMTS system faced a high demand for a very limited channel resource. Moreover, each base-station antenna had to be located on a tall structure and had to transmit at high power in order to provide coverage throughout the entire service area. Because of these high power requirements, all subscriber units in the IMTS system were motor-vehicle-based instruments that carried large storage batteries.

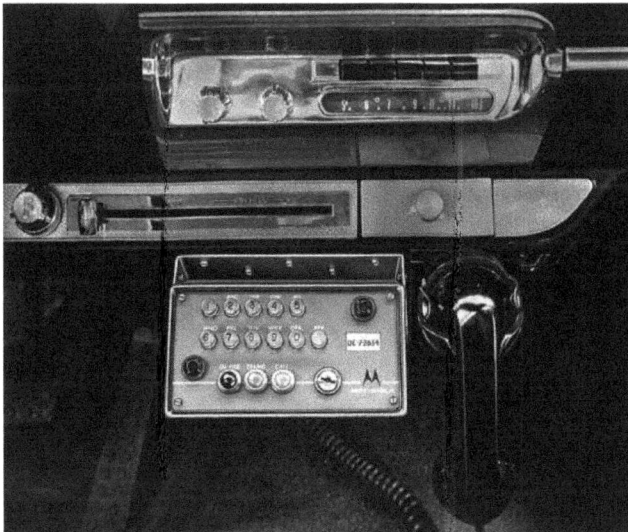

Motorola push-button car telephone, control unit, and handset
mounted under the automobile dashboard.

During this time a truly cellular system, known as the advanced mobile phone system, or AMPS, was developed primarily by AT&T and Motorola, Inc. AMPS was based on 666 paired voice channels, spaced every 30 kilohertz in the

800-megahertz region. The system employed an analog modulation approach frequency modulation, or FM and was designed from the outset to support subscriber units for use both in automobiles and by pedestrians. It was publicly introduced in Chica-go in 1983 and was a success from the beginning. At the end of the first year of service, there were a total of 200,000 AMPS subscribers throughout the United States; five years later there were more than 2,000,000. In response to expected service shortages, the American cellular industry proposed several methods for increasing capacity without requiring additional spectrum allocations. One analog FM approach, proposed by Motorola in 1991, was known as narrowband AMPS, or NAMPS. In NAMPS systems each existing 30-kilohertz voice channel was split into three 10-kilohertz channels. Thus, in place of the 832 channels available in AMPS systems, the NAMPS system offered 2,496 channels. A second approach, developed by a committee of the Telecommunications Industry Association (TIA) in 1988, employed digital modulation and digital voice compression in conjunction with a time-division multiple access (TDMA) method; this also permitted three new voice channels in place of one AMPS channel. Finally, in 1994 there surfaced a third approach, developed originally by Qualcomm, Inc., but also adopted as a standard by the TIA. This third approach used a form of spread spectrum multiple access known as code-division multiple access (CDMA) a technique that, like the original TIA approach, combined digital voice compression with digital modulation. The CDMA system offered 10 to 20 times the capacity of existing AMPS cellular techniques. All of these improved-capacity cellular systems were eventually deployed in the United States, but, since they were incompatible with one another, they supported rather than replaced the older AMPS standard.

The Motorola DynaTAC 8000X was the world's first portable commercial handheld cellular phone.

Although AMPS was the first cellular system to be developed, a Japanese system was the first cellular system to be deployed, in 1979. Other systems that preceded AMPS in operation include the Nordic mobile telephone (NMT) system, deployed in 1981 in Denmark, Finland, Norway, and Sweden, and the total access communication system (TACS), deployed in the United Kingdom in 1983. A number of other cellular systems were developed and deployed in many more countries in the following years. All of them were incompatible with one another. In 1988 a group of government-owned public telephone bodies within the European Community announced the digital global system for mobile communications, referred to as GSM, the first such system that would permit any cellular user in one European country to operate in another European country with the same equipment. GSM soon became ubiquitous throughout Europe.

The analog cellular systems of the 1980s are now referred to as "first-generation" (or 1G) systems, and the digital systems that began to appear in the late 1980s and early '90s are known as the "second generation" (2G). Since the introduction of 2G cell phones, various enhancements have been made in order to provide data services and applications such as Internet browsing, two-way text messaging, still-image transmission, and mobile access by personal computers. One of the most successful applications of this kind is iMode, launched in 1999 in Japan by NTT DoCoMo, the mobile service division of the Nippon Telegraph and Telephone Corporation. Supporting Internet access to selected Web sites, interactive games, information retrieval, and text messaging, iMode became extremely successful; within three years of its introduction, more than 35 million users in Japan had iMode-enabled cell phones.

Motorola's MicroTAC flip cellular phone.

Beginning in 1985, a study group of the Geneva-based International Telecommunication Union (ITU) began to consider specifications for Future Public Land Mobile

Telephone Systems (FPLMTS). These specifications eventually became the basis for a set of "third-generation" (3G) cellular standards, known collectively as IMT-2000. The 3G standards are based loosely on several attributes: the use of CDMA technology; the ability eventually to support three classes of users (vehicle-based, pedestrian, and fixed); and the ability to support voice, data, and multimedia services. The world's first 3G service began in Japan in October 2001 with a system offered by NTT DoCoMo. Soon 3G service was being offered by a number of different carriers in Japan, South Korea, the United States, and other countries. Several new types of service compatible with the higher data rates of 3G systems have become commercially available, including full-motion video transmission, image transmission, location-aware services (through the use of global positioning system [GPS] technology), and high-rate data transmission.

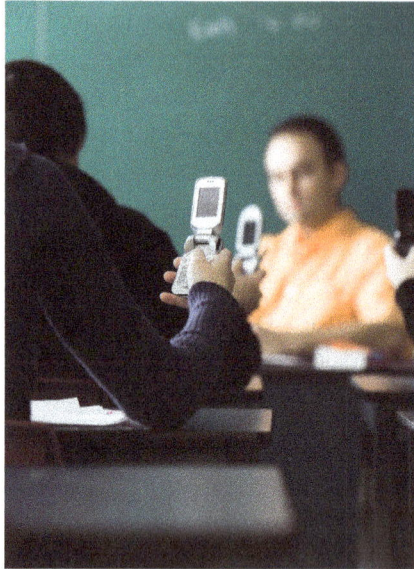

Cell phones became ubiquitous in classrooms around the world for exchanging images and text messages.

The increasing demands placed on mobile telephones to handle even more data than 3G could led to the development of 4G technology. In 2008 the ITU set forward a list of requirements for what it called IMT-Advanced, or 4G; these requirements included data rates of 1 gigabit per second for a stationary user and 100 megabits per second for a moving user. The ITU in 2010 decided that two technologies, LTE-Advanced (Long Term Evolution; LTE) and WirelessMan-Advanced (also called WiMAX), met the requirements. The Swedish telephone company TeliaSonera introduced the first 4G LTE network in Stockholm in 2009.

## Airborne Cellular Systems

In addition to the terrestrial cellular phone systems, there also exist several systems that permit the placement of telephone calls to the PSTN by passengers on commercial aircraft. These in-flight telephones, known by the generic name aeronautical public correspondence (APC) systems, are of two types: terrestrial-based, in which telephone calls are placed directly from an aircraft to an en route ground station; and satellite-based, in which telephone calls are relayed via satellite to a ground station. In the United States the North American terrestrial system (NATS) was introduced by GTE Corporation in 1984. Within a decade the system was installed in more than 1,700 aircraft, with ground stations in the United States providing coverage over most of the United States and southern Canada. A second-generation system, GTE Airfone GenStar, employed digital modulation. In Europe the European Telecommunications Standards Institute (ETSI) adopted a terrestrial APC system known as the terrestrial flight telephone system (TFTS) in 1992. This system employs digital modulation methods and operates in the 1,670–1,675- and 1,800–1,805-megahertz bands. In order to cover most of Europe, the ground stations must be spaced every 50 to 700 km (30 to 435 miles).

# Electromagnetic Spectrum

The electromagnetic spectrum incorporates the range of all electromagnetic radiation, and extends from electric power at the long-wavelength end to gamma radiation at the short-wavelength end. In between, we find radio waves, infra-red, visible light, ultra violet and X-rays used in medical diagnostics. In principle, the spectrum is claimed to be the size of the universe itself but its different parts are limited to certain ranges of electromagnetic waves.

Electromagnetic waves are defined by their special characteristics, such as frequency, wavelength and amplitude. The frequency refers to the number of waves generated in a set period of time and is measured in hertz (Hz). 1 Hz means one wave per second, 1 kHz (kilohertz) means one thousand waves per second, 1 MHz (megahertz) means one million waves per second, 1GHz (gigahertz) means one billion waves per second and so on.

Wavelength is the distance between two waves. There is a fixed mathematical interrelation between the frequency and the wavelength. The higher frequencies have shorter wavelengths and the lower frequencies have longer wavelengths. The wavelength also indicates the ability of the wave to travel in space. A lower frequency wave can reach longer distances than a higher frequency wave. Radio waves are usually specified by frequency rather than wavelength.

The radio frequency spectrum (which is simply referred to as spectrum) is only a comparatively small part of the electromagnetic spectrum, covering the range from 3 Hz to 300 GHz. It includes a range of a certain type of electromagnetic waves, called the radio waves, generated by transmitters and received by antennas or aerials.

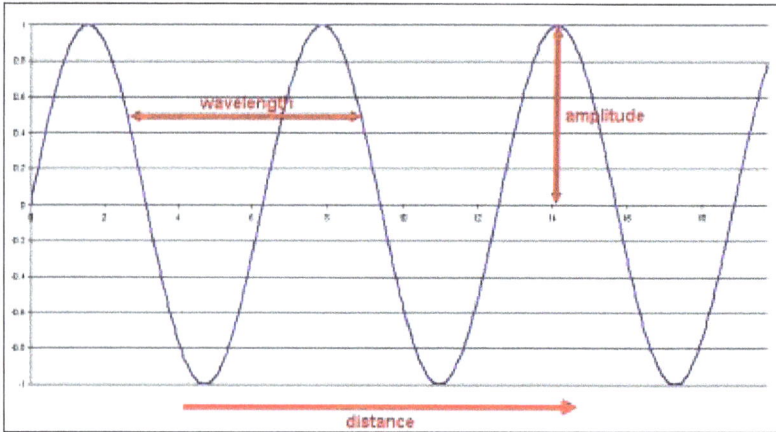

Electromagnetic wave.

## How Radio Spectrum Works?

The radio spectrum is the home of communication technologies such as mobile phones, radio and television broadcasting, two-way radios, broadband services, radar, fixed links, satellite communications, etc. due to its excellent ability to carry codified information (signals). It is relatively cheap to build the infrastructure which can also provide mobility and portability.

Depending on the frequency range, the radio spectrum is divided into frequency bands and sub-bands, as illustrated in figure.

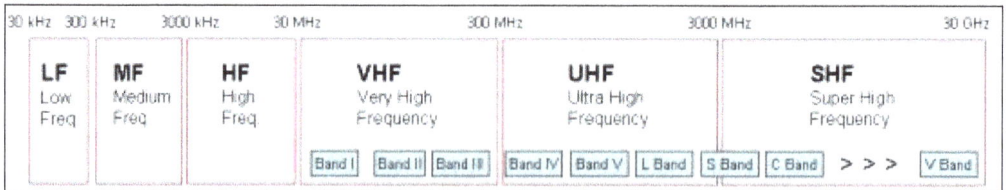

Frequency bands and sub-bands.

In theory, different communication technologies could exist in any part of the radio spectrum, but the more information a signal is to carry, the more bandwidth it needs. In simple terms, bandwidth is the range of frequencies that a signal occupies in spectrum. For example, an FM radio station might broadcast on the 92.9 MHz frequency, but requires 0.3 MHz (equivalent of 300 kHz) bandwidth – the spectrum between the frequencies 92.8 and 93.0 MHz inclusive. Other stations cannot broadcast on these frequencies within the same area without causing or receiving interference.

For planning purposes, the spectrum bands are divided into channels. The bandwidth of spectrum channels can vary band by band. VHF Band II, the home of FM radio, for instance, is sliced up in 100 kHz-wide channels. An FM station requires 300 kHz bandwidth, therefore each FM radio station takes up three spectrum channels. In the case of television broadcasting, the agreed bandwidth of a channel is 8 MHz in UHF Band IV/V. The bandwidth requirement of an analogue TV programme channel happens to be the same as the bandwidth of one spectrum TV channel, i.e. 8 MHz.

Lower frequencies have less bandwidth capacity than higher frequencies. It means that signals that carry a lot of information (such as television, broadband or mobile phones) are better placed in the higher frequency bands while simple radio (audio) signals can be carried by the low frequency waves. Since low frequencies travel long distances but have less bandwidth capacity, placing one television channel (which uses a lot of bandwidth) in the UK in the lower frequency bands would mean that most of the Long Wave and Medium Wave radio services from Northern Europe to Sub-Saharan Africa would be squeezed out.

Once a radio signal has been transmitted, it has certain propagation characteristics associated with its frequency. Propagation describes the behaviour of a radio wave in spectrum. In different bands, waves have distinct abilities to hop, spread and penetrate. Certain waves can go through or bounce off walls or curve around corners better than others. Your mobile phone will probably work inside a building because its signal goes through windows, but you will generally need a rooftop aerial for your TV set to achieve good reception. Figure describes the propagation characteristics of the radio frequency bands.

| Frequency Band | Propagation mode (the way radio waves spread in spectrum) | Coverage |
| --- | --- | --- |
| Very Low Frequency | On the ground | Long distances, e.g. for submarine communications and time code signals. |
| Low Frequency | On the ground and in the sky at night | Country wide. Some reduction of coverage at night due to reflections from the ionosphere. |
| Medium Frequency | On the ground and in the sky at night | Regions of a country. At night time, coverage is significantly reduced by signals reflected from the ionosphere. |
| High Frequency | Hopping between the ground and the sky | Long distance coverage to continents. A range of High Frequencies are needed to provide continuous coverage during the day and night and at different times of the year. |
| Very High Frequency | In line-of-sight, but for short periods, the wave enters the troposphere (the lowermost part of the Earth's atmosphere) | High power broadcasting stations provide coverage up to around 50 to 70 km radius. For short periods of time signals can propagate for long distances in the troposphere (the lowermost portion of the Earth's atmosphere) and cause interference between services on the same frequency. |

| Ultra High Frequency | In line-of-sight, and tropospheric for short periods | Similar range to VHF but requires many more filler stations to overcome obstructions to the signal arising from the attenuation of terrain features. |
|---|---|---|
| Super High Frequency | Between focussed points and in line-ofsight | Need a clear line-of-sight path as signals blocked by buildings or other objects. Ideally suited for satellite communications and fixed links where highly focused antennas (dishes) can be used or for short range coverage, e.g. inside buildings. |
| Extremely High Frequency | Between very focussed points and in line-ofsight | Short paths and no possibility for penetrating building walls. |

In order to understand how radio spectrum works, one more buzzword has to be remembered: modulation. Modulation is the actual process of encoding information in a radio signal by varying the characteristics (the amplitude, the frequency or the phase) of the radio wave. Simple examples of the resulting waves are illustrated in figure.

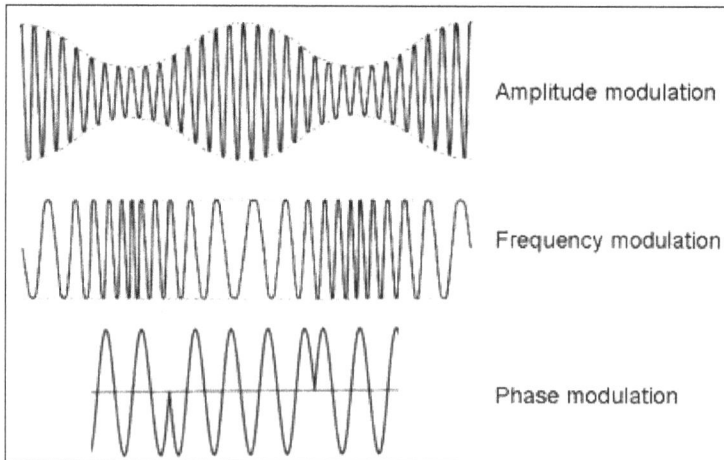

Types of radio wave modulation.

Amplitude modulation (AM) has been used to generate carrier waves for AM radio stations which cover large areas. Radio 4 on long wave (LW), for instance, is carried by an amplitude modulated signal. Frequency modulation (FM) is used for FM broadcasting which provides better sound quality to AM radio but the signal does not travel as far as an AM signal.

Phase modulation (PM) and amplitude modulation is used to encode digital information (consisting of 0s and 1s) into radio signals. There are very complex advanced variants of these modulation techniques which allow for large amounts of digital data to be encoded or compressed into a signal.

## International Harmonisation

Radio waves do not respect international borders. Signals can cross boundaries easily. International harmonisation – to reduce the scope for unwelcome interference between one country and another – takes place at three levels:

- The International Telecommunications Union on a worldwide basis.

- The Conference of Postal and Telecommunications Administration (which brings together 46 countries) in Europe.

- A bilateral country-by-country basis (for example, to ensure that transmitters in the South of England do not cause interference in France and ensure that French transmissions do not interfere with services in the South of England).

International harmonisation of spectrum bands for particular uses helps create valuable economies of scale. The scope to use mobile phones across the European Union, because of such harmonisation, improves the market for consumers. And harmonisation provides the prospect of a mass market, and lower prices, for receiving equipment.

A major international planning conference (often referred to as RRC06) in the spring of 2006 agreed a harmonised plan (GE06) for digital terrestrial broadcasting in Bands III, IV and V for Europe, Africa and many other countries. Almost all of the spectrum requirements of each country were met.

## Spectrum Availability

Spectrum that can be used in new and innovative ways is regularly becoming available as new technologies make more efficient use of the spectrum and obsolete technologies free up spectrum space.

Change is taking place in various frequency bands, although in some cases, analogue and digital technologies will co-exist for quite some time.

The Low Frequency (LF), Medium Frequency (MF) and High Frequency (HF) broadcasting bands (below 30 MHz) are still used in much the same way as they always have been since the birth of radio broadcasting over 80 years ago for Long Wave (LW), Medium Wave (MW) and Short Wave (SW) analogue broadcasting. BBC Radio 4 is still being broadcast on LW and BBC World Service programmes are distributed on SW in the HF band. But, also in the HF band, a growing number of transmissions are being established in digital (DRM) format, primarily for international broadcasting. In the MF band, a limited range of frequencies are available for local analogue Medium Wave (MW) radio services.

A part of the Very High Frequency (VHF) band is used intensively for FM sound broadcasting in most countries and planning of new analogue services is still being carried out. There are a limited number of frequencies available for regional, local and community stations. Currently the re-allocation of this spectrum for digital services is difficult to envisage. In the longer term, digital services such as DRM+ Digital Radio could use this band, but the technology has not yet been fully tested.

The ongoing debate about spectrum availability in the UK is focussing on a "sweetspot" where most modern communication technologies such as DAB Digital Radio, digital television, 3G mobile phones and WiFi wireless Internet access services operate. The sweetspot, in fact, is the upper part of the Very High Frequency (VHF) band and the whole of the Ultra High Frequency (UHF) band, incorporating frequencies from around 200 MHz to 3 GHz as illustrated in figure.

The top end of the VHF band (known as Band III) is used for DAB Digital Radio Broadcasting. A total of seven frequency blocks are currently used here for two national and 46 local and regional DAB multiplexes. Four additional frequency blocks have been recently advertised for licensing.

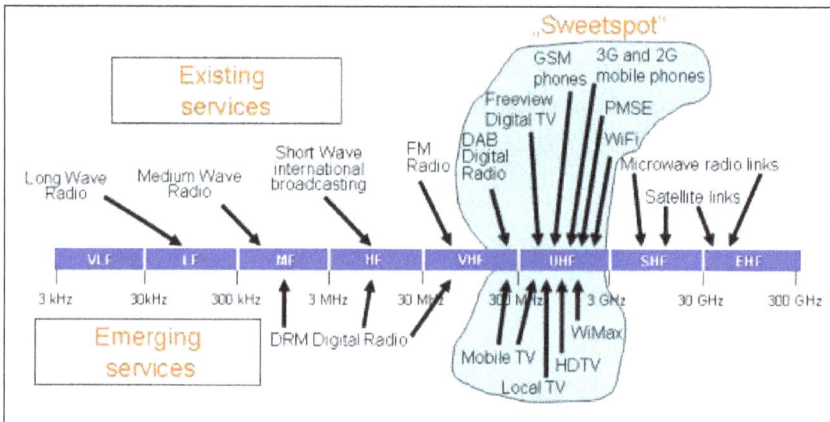

The "sweetspot" in the radio spectrum.

The UHF band includes four named sub-bands: Band IV, Band V, Lband and S-band as shown in figure. These sub-bands also differ from each other in certain characteristics and uses are not necessarily interchangeable between them.

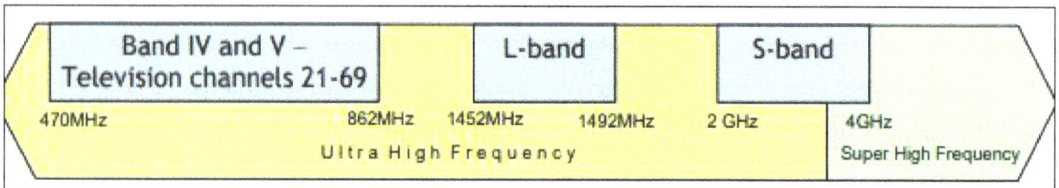

Sub-bands in the Ultra High Frequency Band.

UHF Band IV/V is divided into 49 channels. 46 of them are currently used for both analogue and digital television broadcasting. After digital switchover, the six existing television multiplexes will occupy 32 channels. The "digital dividend", the spectrum to be afforded by analogue switch-off will be equivalent to 14 spectrum television channels, each containing 8 MHz bandwidth. The total spectrum becoming available during the digital switchover process from 2008 through to 2012 will be 14 x 8 MHz = 112 MHz. Figure shows what will become available in Band IV/V after digital switchover.

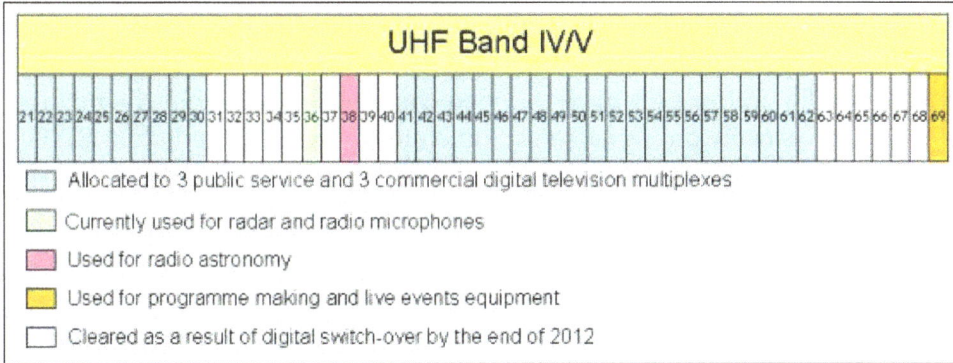

Spectrum availability in UHF Band IV/V.

Public attention heavily focuses on the "digital dividend" as it can host a number of new and innovative services such as High Definition Television, mobile TV or broadband wireless access services. OFCOM is currently undertaking research to define the possible uses of the "digital dividend" and is examining options to make some of the released spectrum available for other uses on a rolling basis region by region from 2008, the start of the switch-over process.

Some other parts of the spectrum will become available sooner. The Lband, is an interesting possibility for multimedia services such as mobile TV or wireless internet access as there would be scope for harmonisation of this band at pan European level.

UHF Channel 36 (currently used for radar and radio microphones) is being considered to be released for other uses. Potential contenders for this spectrum could be mobile TV, broadband wireless access and terrestrial digital broadcast services.

## New and Emerging Technologies Explained

There are many services that could use newly available spectrum. Before considering deployment options, it might be useful to describe the technologies that underpin them.

Digital terrestrial television (DTT): DTT services are broadcast by multiplexes that encode picture and sound of several TV programme channels and some interactive information services in one signal. The signal is then decoded by either an Integrated Digital Television (IDTV) set or a set-top box (e.g. Freeview) connected to the TV set. The technology is called Digital Video Broadcasting Terrestrial (DVB-T).

There are three public service and three commercial multiplexes serving the UK. Just like an analogue television channel, a multiplex requires 8 MHz-wide spectrum channels. However, using just one 8 MHz-wide channel, several Standard Definition

Television (SDTV) programme channels can be provided to the public on each of the six multiplexes. This is what you see on your screen today, for instance, through your Freeview box.

The precise number of TV channels in a multiplex depends on the transmission mode employed, the range of services to be accommodated and the desired level of picture quality. Some DTT multiplexes have 10 video channels in them, but the pictures look less than ideal, especially on big screens. Some multiplexes host only four SDTV channels and the pictures look great.

High Definition Television (HDTV), which has not yet been introduced on Freeview, offers an enhanced viewing experience with sharper picture quality and improved sound. It is ideal for over 30-inch big screens. Importantly, HDTV is much more demanding of spectrum – it could take up nearly three times the data transmission capacity requirement of SDTV.

HDTV services are already available in many countries, including Japan, South Korea, France and USA. Most HDTV channels are carried on satellite platforms at the moment and some of them are provided as a premium service. Free-to-air broadcasters are seeking to launch their HDTV services on terrestrial platforms such as Freeview.

Local television: There is growing interest among companies in providing local TV services, either across a conurbation or in much localized areas.

Local TV services could have two technical options for delivery on digital terrestrial broadcasting platforms.

First, they might acquire space on a regional multiplex transmitter. National DTT networks use a number of regional transmitter sites to achieve UK-wide coverage and some spare capacity might be available on some of these. However, so far these multiplexes have tended to operate at full capacity and the scope for regional "add ons" to serve local TV interests may not be great.

The second option is to take advantage of "interleaved spectrum" which might be available in the area they wish to cover. "Interleaved spectrum" is the by-product of national networks that use several 8 MHz-wide channels in the UHF Band IV/V to cover the UK. Some of these channels might be not used in certain areas and could be allocated to low-power local TV multiplexes. One interleaved channel could provide two programme channels. The signal would be sufficiently rugged in the local area and would be automatically picked up by Freeview boxes.

- Mobile telephony: Mobile phones, the most successful communications development of our times, occupy various parts of the spectrum. Thus, "2G" (Second Generation) phones – the standard mobile phone – operate just under the 1GHz and around the 1.75 GHz band. The increasingly popular "3G" (Third

Generation) phones operate around the 2GHz band. In addition, mobile expansion bands have been allocated in other parts of the spectrum to allow the delivery of additional services by mobile.

- Mobile TV: A potentially important innovative service, mobile TV would enable users to watch TV wherever they want. Already there are several different standards which will be vying for market success. For instance:

  ○ Mobile TV on DAB-IP (like BT Movio) has already been tried in the UK for Windows Mobile-based Smartphones. These phones are enabled with Digital Audio Broadcasting Internet Protocol (DAB-IP) technology, using "3G" mobiles as platform.

  ○ "3G" mobiles themselves might provide a platform for mobile TV services, though transmission capacity problems could be formidable. The mobile operator Hutchinson 3G UK Ltd. has already made TV services available on mobile phones.

  ○ Digital Video Broadcasting Handheld (DVB-H) is a more robust means of getting TV content to the user. It might provide up to 20 programme channels in an 8 MHz spectrum channel.

  ○ Qualcomm MediaFLO is an American standard for mobile TV that is being tested by BSkyB in the UK. It works similarly to DVB-H.

  ○ Digital Multimedia Broadcasting Terrestrial (DMB-T), delivered on the DAB Digital Radio platform, can also provide mobile TV.

  ○ It might sound a bit confusing that digital radio platforms can deliver television or video services. But the nature of the digital signal is that it can carry practically any information on any platform if the receiver is designed to process that information.

- WiMax (Worldwide Interoperability for Microwave Access): Regarded as a revolutionary technology for internet wireless access, WiMax, in theory, could provide the service up to 30 miles from the base station. The technology has several standards. The latest one is designed for a theoretical connection speed of up to 75 megabits per second. WiMax can utilise a wide range of spectrum bands from 2 to 66 GHz and can take up channels of varying bandwidth from 1.25 MHz to 20 MHz.

  Currently there is no standard for WiMax below 1GHz but it is likely that one will be developed soon as this is seen to be attractive for providing services to rural locations.

- DAB Digital Radio: Launched by the BBC in 1995, DAB Digital Radio is now a rapidly growing service in the UK. Several technical standards exist for this technology, but in Europe, the Eureka 147 T-DAB system has prevailed. For listeners, it means that a digital radio set purchased in the UK can also be used in any other country where digital radio is provided in the Eureka 147 format.

These services are broadcast by T-DAB terrestrial multiplexes – transmitters that combine a number of radio stations in one signal. There are two national (BBC and Digital One) and a number of local multiplexes in the UK.

Each multiplex occupies 1.7 MHz of bandwidth and the use of single frequency network (SFN) configurations means that only one frequency block is necessary to provide UK-wide coverage of the BBC's national radio services. The number of radio stations that a multiplex can carry depends on the sound quality expected of each station. As a general rule, around nine radio stations at a joint stereo bit rate of 128 kilobit per second can be accommodated on a single multiplex. DAB can also carry video channels for small handheld screens.

- Digital Radio Mondiale (DRM) is the digital technology for Short Wave (SW) international radio broadcasting in the High Frequency band, but trials have also taken place for the Medium Frequency band. SW signals can reach large distances but reception is usually poor.

DRM offers a dramatic enhancement in sound quality, and mitigates the effects of audible interference from other stations. It is also designed to make receiver operation more user-friendly. DRM promises to re-invigorate the use of the Low, Medium and High Frequency Bands. New digital radio receivers are under development to combine both DAB and DRM reception, so one radio set would be capable to pick up local digital radio signals as well as signals coming from Italy, Mexico or even China.

Another version of this technology, called DRM+ is under development for VHF Band II which has traditionally been used by FM radio.

- PMSE: The abbreviation stands for Programme Making and Special Events equipment that are used at concerts, theatres, and filming, recording and live broadcasts. These include cordless microphones, cameras and other cordless devices.

These devices can operate in various spectrum bands and can be interleaved between existing other services due to their low radiated power, thus making efficient use of the spectrum. Their signal reaches just a few meters, with very little chance to interfere with other similar devices. However, they still need their well-defined spectrum space so that other technologies do not interfere with them.

Further into the future, Software Defined Radio (SDR) and Cognitive Radio (CR) might be very attractive both to users and spectrum planners. These are not radio sets, but technologies that would combine several services that use radio waves. SDR users would simply request a service through the device which would then negotiate with the network to identify the most appropriate frequency for that service.

Cognitive radio would have the additional ability to recognize and distinguish signals, making spectrum practically abundant. Again, this technology is still in infancy.

## Competition or Co-habitation

Just as certain types of plants are best grown on particular types of soil, not all technologies are suited to all frequency ranges. Certain services may be more suitable for particular frequency bands. This may be because:

- Different services have different needs. Broadcasting, for instance, is a one-way communication: The transmitter sends a signal to the receiver. Mobile phones or WiFi devices have to "talk back" to the base station to upload as well as download information, so they need frequencies to enable this two-way communication to take place.

- The propagation is different in each frequency band. Higher frequencies can provide more rugged signals for mobile communication devices than lower frequencies. Mobile phones usually work on trains or inside buildings due to the construction of dense base station networks which are needed to provide the link from the low power mobile phone to the base station. Try to use an FM radio on a train; it probably won't work very well because the metal structure of the carriage blocks the FM signal.

- Different constraints exist on transmitter and receiver equipment design. Bigger antennas are needed to receive the signal on lower frequencies while higher frequency signals can be detected by smaller antennas. Think of your FM kitchen radio or your HiFi set at home which needs a fairly long antenna (sometimes the rooftop aerial has to be plugged into the HiFi) to get good reception. Early GSM mobile phones also needed extendable antennas. Your 2G or 3G mobile phone, on the other hand, operates with a very small antenna; you can't even see it as it is hidden inside the phone.

- Moving a service from one band to another might require users to retune or to change the receiving device. This could undermine the sustainability of the service given the vast quantities of TV and radio receivers in people's homes.

- Different international co-ordination puts constraints on different bands.

These considerations influence the way in which different technologies are deployed.

Nevertheless, some technologies have more possible outlets than others. Figure indicates the different bands that could, in theory, be used to deliver a range of services.

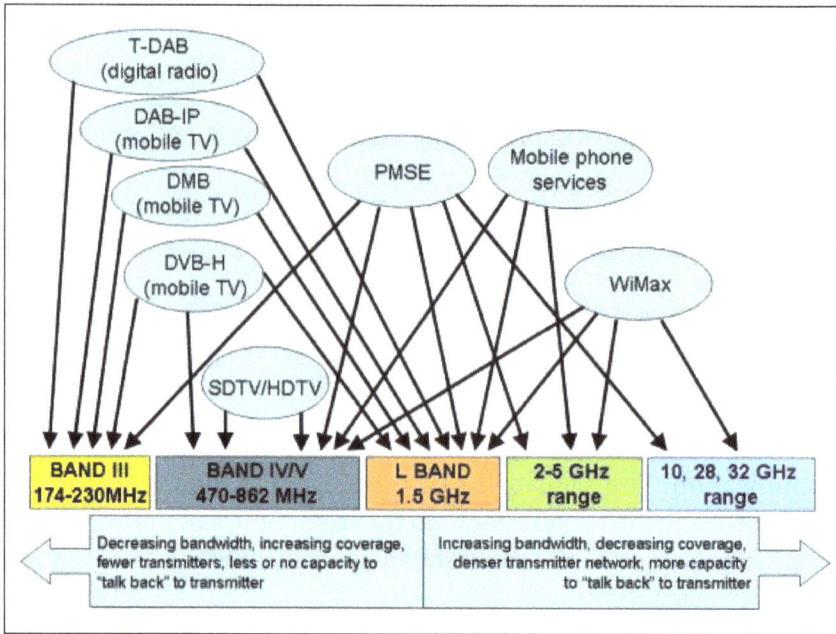

Alternative frequency bands for digital technologies.

Mobile TV technologies can be deployed in several bands. DAB-based services have been optimised for Band III or the L-band, while DVB-H is designed to operate in Band III, IV and V or even the L-band. Companies wishing to provide such services will have to examine their options carefully regarding both the technologies and the bands. Acquisition of spectrum in more heavily used bands, like Band IV/V could prove too costly to make it an affordable service.

DAB-based mobile TV services can co-exist with radio services on national, regional and local T-DAB multiplexes. They operate in Band III at present, and some capacity might be available for mobile TV on the existing multiplexes.

Some DVB-H mobile TV services could be accommodated in Band IV/V. For example, Channel 36 (currently used for radar and radio microphones) could be assigned to mobile TV as well as a few other channels which will be available as part of the "digital dividend" after digital switch-over. Channel 36, however, might be problematic to co-ordinate with neighbouring countries as they might also seek to introduce new high-power assignments in that channel, limiting its use within the UK. Television broadcasters might also be strong contenders for these channels as SDTV and HDTV services have no other deployment options outside Band IV/V.

Local TV providers could put further demand on Band IV/V as they can operate in the spectrum interleaved regionally between the channels used by national DVB-T television multiplexes. But interleaved spectrum could also be used to enable Programme

Making and Special Events (PMSE) equipment and WiMax broadband wireless access services to operate in Band IV/V.

WiMax providers might seek to secure channels in Band IV/V for broadband wireless access services. Here, however, there has to be a tradeoff between how many users can be supported in a cell, the available data transfer rate and the number of network providers. Although the use of Band IV/V could make the coverage area (the cell size) bigger, due to these constraints, channels in this band might only be required as a way of delivering WiMax services to remote rural communities where the number of users per cell could be relatively small. Indeed, the proponents of WiMax systems seem to be favouring higher frequencies such as the 2.5 and 3.5 GHz or the 10, 28 and 32 GHz bands (the latter bands only support short range indoor reception).

Mobile phone services could also use Band IV/V and some service providers are interested in this band, particularly for providing coverage in rural areas. However, there are compatibility issues concerning sharing with broadcasting services and these would need to be studied. There is also a 190 MHz expansion band at 2.5 GHz which is harmonised throughout Europe for mobile phone services.

The L-band (also known as the 1.5 GHz band) can support a number of different approaches. Propagation in this band could provide better conditions for mobile users. Radio signals in L-band go through windows and can benefit from reflections, particularly in built up areas, so they can reach receivers "on the move" (on trains, buses etc.). But the networks would require more transmitters therefore infrastructure could involve higher costs.

Current European frequency plans harmonise the use of the L-band for DAB technology which supports both radio and mobile TV services. In the UK, T-DAB is presently placed in Band III, but several countries including France, Germany and the Czech Republic operate T-DAB in the 1.5 GHz band, although this has not yet proved to be a great success.

Regulators, however, seem to be open to the idea to change international harmonisation rules in the future and allow technologies such as DVB-H, DMB-T (mobile TV) and WiMax to use this band. Further technical research might be necessary to establish the feasibility of these services in the L-band.

Competition for spectrum seems to be inevitable as market players try to capture opportunities to launch new services. Alternative deployment solutions for various technologies might ease the demand for spectrum in certain frequency bands. Band IV/V could offer more spectrum after digital switch-over than any other band but the demand for additional SDTV and HDTV services could be high. The level of consumer demand and viable business models would have to be established for new services such as mobile TV and broadband wireless access before their need for spectrum can be assessed with confidence. Some degree of planning could mitigate these uncertainties and encourage the development and co-existence of innovative services.

## Radio Waves

A radio wave is a type of electromagnetic signal designed to carry information through the air over relatively long distances. Sometimes radio waves are referred to as radio frequency (RF) signals. These signals oscillate at a very high frequency, which allows the waves to travel through the air similar to waves on an ocean. Radio waves have been in use for many years. They provide the means for carrying music to FM radios and video to televisions. In addition, radio waves are the primary means for carrying data over a wireless network. As shown in figure, a radio wave has amplitude, frequency, and phase elements. These attributes may be varied in time to represent information.

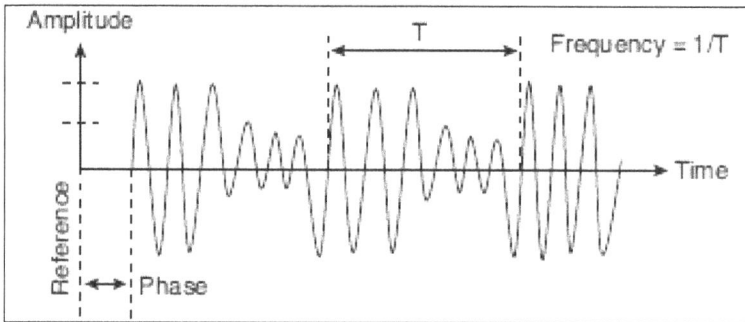

The amplitude frequency and phase elements of a radio wave.

### Amplitude

The amplitude of a radio wave indicates its strength. The measure for amplitude is generally power, which is analogous to the amount of effort a person needs to exert to ride a bicycle over a specific distance. Similarly, power in terms of electromagnetic signals represents the amount of energy necessary to push the signal over a particular distance. As the power increases, so does the range.

Radio waves have amplitudes with units of watts, which represent the amount of power in the signal. Watts have linear characteristics that follow mathematical relationships we are all very familiar with. For example, the result of doubling 10 milliwatts (mW) is 20 mW. We certainly do not need to do any serious number crunching to get that result.

As an alternative, it is possible to use dBm units (decibels referenced to 1 mW) to represent the amplitude of radio waves. The dBm is the amount of power in watts referenced to 1 mW. Zero (0) dBm equals 1 mW. By the way, the little m in dBm is a good reminder of the 1 mW reference. The dBm values are positive above 1 mW and negative below 1 mW. Beyond that, math with dBm values gets a bit harder.

You can adjust the transmit power of most client cards and access points. For example, some access points allow you to set the transmit power in increments from −1 dBm (0.78 mW) up to 23 dBm (200 mW).

## Frequency

The frequency of a radio wave is the number of times per second that the signal repeats itself. The unit for frequency is Hertz (Hz), which is actually the number of cycles occurring each second. In fact, an old convention for the unit for frequency is cycles per second (cps).

802.11 WLANs use radio waves having frequencies of 2.4 GHz and 5 GHz, which means that the signal includes 2,400,000,000 cycles per second and 5,000,000,000 cycles per second, respectively. Signals operating at these frequencies are too high for humans to hear and too low for humans to see. Thus, radio waves are not noticed by humans.

The frequency impacts the propagation of radio waves. Theoretically, higher-frequency signals propagate over a shorter range than lower-frequency signals. In practice, however, the range of different frequency signals might be the same or higher-frequency signals might propagate farther than lower-frequency signals. For example, a 5-GHz signal transmitted at a higher transmit power might go farther than a 2.4-GHz signal transmitted at a lower power, especially if electrical noise in the area impacts the 5-GHz part of the radio spectrum less than the 2.4-GHz portion of the spectrum.

## Phase

The phase of a radio wave corresponds to how far the signal is offset from a reference point (such as a particular time or another signal). By convention, each cycle of the signal spans 360 degrees. For example, a signal might have a phase shift of 90 degrees, which means that the offset amount is one-quarter (90/360 = 1/4) of the signal.

## RF System Components

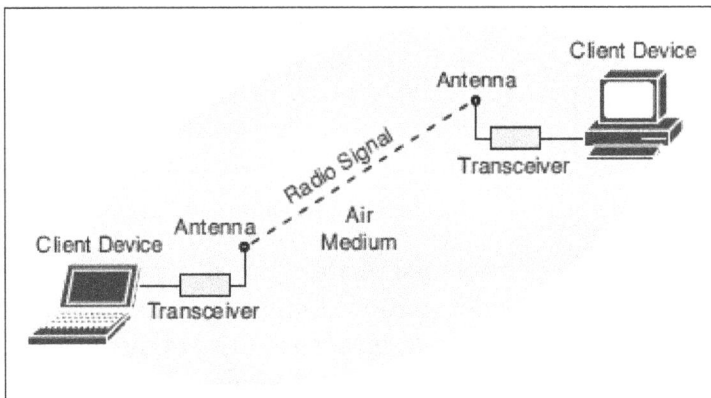

An Rf system consists of RF transceivers, and a transmission medium.

Figure illustrates a basic RF system that enables the propagation of radio waves. The transceiver and antenna can be integrated inside the client device or can be an external component. The transmission medium is primarily air, but there might be obstacles, such as walls and furniture.

## RF Transceiver

A key component of a WLAN is the RF transceiver, which consists of a transmitter and a receiver. The transmitter transmits the radio wave on one end of the system (the "source"), and the receiver receives the radio wave on the other side (the "destination") of the system. The transceiver is generally composed of hardware that is part of the wireless client radio device (sometimes referred to as a client card).

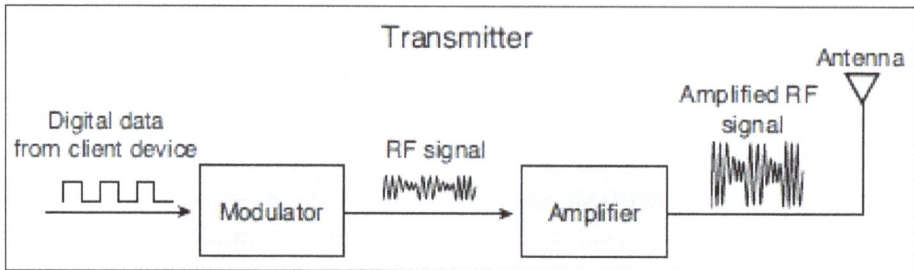

A transmitter consists of a modulator, an amplifier and an antenna.

Figure shows the basic components of a transmitter. A process known as *modulation* converts electrical digital signals that represent information (data bits, 1s and 0s) inside a computer into radio waves at the desired frequency, which propagate through the air medium. The amplifier increases the amplitude of the radio wave signal to a desired transmit power prior to being fed to the antenna and propagating through the transmission medium (consisting primarily of air in addition to obstacles, such as walls, ceilings, chairs, and so on).

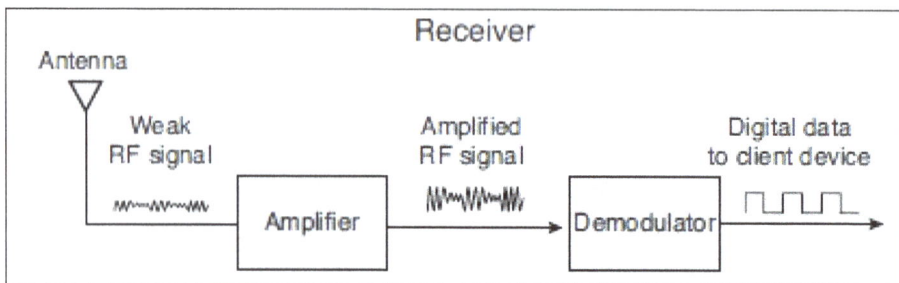

A receiver consists of an antenna an amplifier and a demodulator.

At the destination, a receiver detects the relatively weak RF signal and demodulates it into data types applicable to the destination computer. The radio wave at the receiver must have amplitude that is above the receiver sensitivity of the receiver; otherwise, the receiver will not be able to "interpret" the signal, or decode it. The minimum receiver sensitivity depends on the data rate. For example, say that the receiver sensitivity of an access point is −69 dBm for 300 Mbps (802.11n) and −90 dBm for 1 Mbps (802.11b). The amplitude of the radio wave at the receiver of this access point must be above −69 dBm for 300 Mbps or above −90 dBm for 1 Mbps before the receiver will be able to decode the signal.

## RF Modulation

RF modulation transforms digital data, such as binary 1s and 0s representing an e-mail message, from the network into an RF signal suitable for transmission through the air. This involves converting the digital signal representing the data into an analog signal. As part of this process, modulation superimposes the digital data signal onto a carrier signal, which is a radio wave having a specific frequency. In effect, the data rides on top of the carrier. To represent the data, the modulation signal varies the carrier signal in a manner that represents the data.

Modulation is necessary because it is not practical to transmit data in its native form. For example, say that Kimberlyn wants to transmit her voice wirelessly from Dayton to Cincinnati, which is about 65 miles. One approach is for Kimberlyn to use a really high-powered audio amplifier system to boost her voice enough to be heard over a 65-mile range. The problem with this, of course, is that the intense volume would probably deafen everyone in Dayton and all the communities between Dayton and Cincinnati. Instead, a better approach is to modulate Kimberlyn's voice with a radio wave or light carrier signal that's out of range of human hearing and suitable for propagation through the air. The data signal can vary the amplitude, frequency, or phase of the carrier signal, and amplification of the carrier will not bother humans because it is well beyond the hearing range.

The latter is precisely what modulation does. A modulator mixes the source data signal with a carrier signal. In addition, the transmitter couples the resulting modulated and amplified signals to an antenna, which is designed to interface the signal to the air. The modulated signal then departs the antenna and propagates through the air. The receiving station antenna couples the modulated signal into a demodulator, which derives the data signal from the signal carrier.

## Amplitude-shift Keying

One of the simplest forms of modulation is amplitude modulation (sometimes referred to as amplitude-shift keying), which varies the amplitude of a signal to represent data. Figure illustrates this concept. Frequency-shift keying (FSK) is common for lightbased systems whereby the presence of a 1 data bit turns the light on and the presence of a 0 bit turns the light off. Actual light signal codes are more complex, but the main idea is to turn the light on and off to send the data. This is similar to giving flashlights to two people in a dark room and having them communicate with each other by flicking the flashlights on and off to send coded information.

Amplitude modulation alone does not work very well with RF systems because there are signals (noise) present inside buildings and outdoors that alter the amplitude of the radio wave, which causes the receiver to demodulate the signal incorrectly. These noise signals can cause the signal amplitude to be artificially high for a period of time; for example, the receiver would demodulate the signal into something that does not represent what was intended (for example, 10000001101101 would become 10111101101101). To combat impacts from noise, modulation for RF systems is more complex than using only amplitude modulation.

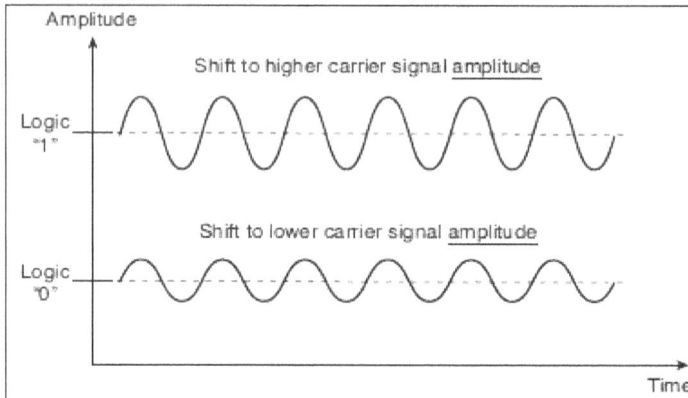

Amplitude-sbift keying varies the amplitude of the signal to represent digital data.

## Frequency-shift Keying

FSK makes slight changes to the frequency of the carrier signal to represent data in a manner that's suitable for propagation through the air at low to moderate data rates. For example, as shown in figure, modulation can represent a 1 or 0 data bit with either a positive or negative shift in frequency of the carrier. If the shift in frequency is negative that is, a shift of the carrier to a lower frequency the result is a logic 0. The receiver can detect this shift in frequency and demodulate the results as a 0 data bit. As a result, FSK avoids the impacts of common noise that exhibits shifts in amplitude.

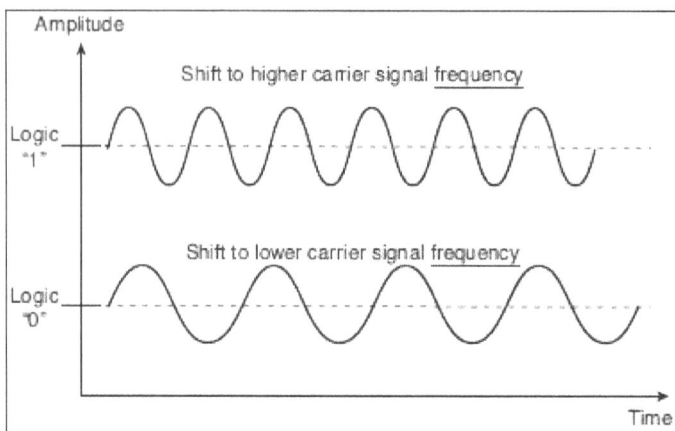

Frequency-sbift keying makes use of changes in frequency to represent digital data.

## Phase-shift Keying

Some systems use phase-shift keying (PSK), which is similar to FSK, for modulation purposes for low to moderate data rates. With PSK, data causes changes in the signal's phase, while the frequency remains constant. The phase shift, as figure depicts, can correspond to a specific positive or negative amount relative to a reference. A receiver can detect these phase shifts and realize the corresponding data bits. As with FSK, PSK is mostly immune to common noise that is based on shifts in amplitude.

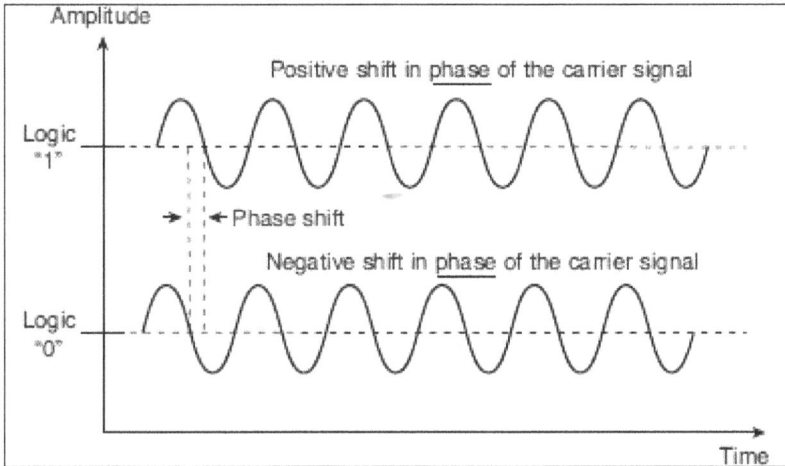

Phase-sbift keying makes use of changes in phase to represent digital data.

## Quadrature Amplitude Modulation

Quadrature amplitude modulation (QAM) causes both the amplitude and phase of the carrier to change to represent patterns of data, often referred to as symbols. The advantage of QAM is the capability of representing large groups of bits as a single amplitude and phase combination. In fact, some QAM-based systems, for example, make use of 64 different phase and amplitude combinations, resulting in the representation of 6 data bits per symbol. Higher-order combinations of phase and amplitude in QAM make it possible for standards such as 802.11n and 802.11ac to support higher data rates.

## Spread Spectrum

After modulating the digital signal into an analog carrier signal using FSK, PSK, or QAM, some WLAN transceivers spread the modulated carrier over a wider spectrum to comply with regulatory rules. This process, called *spread spectrum,* significantly reduces the possibility of outward and inward interference. As a result, regulatory bodies generally do not require users of spread spectrum systems to obtain licenses. Spread spectrum, developed originally by the military, spreads a signal's power over a wide band of frequencies.

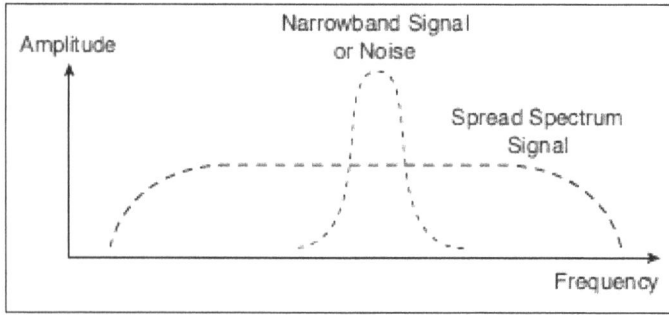

Spread spectrum occupies a wide portion of the RF spectrum.

Spread spectrum radio components use either direct sequence or frequency hopping for spreading the signal. Direct sequence modulates a radio carrier by a digital code with a bit rate much higher than the information signal bandwidth. Figure is a hypothetical example of direct sequence that represents the transmission of three data bits (101) serially in time. The actual transmission is based on a different code word that represents each type of data bit (1 and 0). As shown in the figure, when sending a data bit 1, the radio sends the code word 00010011100 to represent the data bit. Similarly, when sending a data bit 0, the radio sends the code word 11101100011. The increase in the number of bits sent that represents the data effectively spreads the signal across a wider portion of the frequency spectrum.

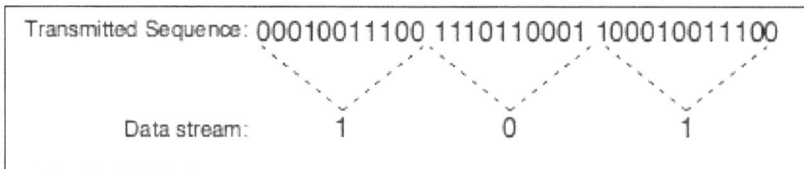

Direct sequence is a type of spread spectrum.

Frequency hopping uses a different technique to spread the signal by quickly hopping the radio carrier from one frequency to another within a specific range. Figure illustrates this concept. The boxes labeled A, B, C, D, and E in the figure represent bursts of data that are sent at different times and frequencies. This also effectively spreads the signal across a wider part of the spectrum.

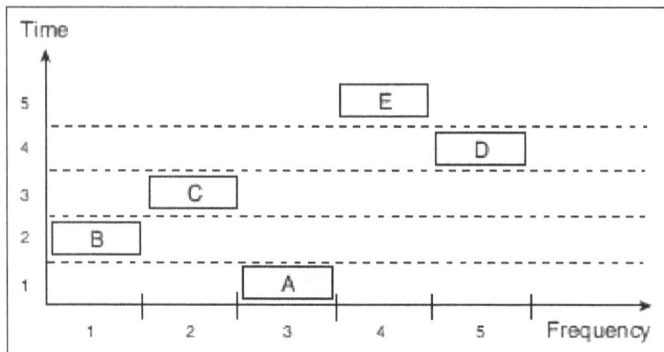

Frequency hopping is a type of spread spectrum.

## Orthogonal Frequency-division Multiplexing

Instead of using spread spectrum, higher-speed WLANs make use of orthogonal frequency- division multiplexing (OFDM). OFDM divides a signal modulated with FSK, PSK, or QAM across multiple subcarriers occupying a specific channel. OFDM is extremely efficient, which enables it to provide the higher data rates and minimize multipath propagation problems. OFDM has also been around for a while, supporting the global standard for asymmetric digital subscriber line (ADSL), a high-speed wired telephony standard.

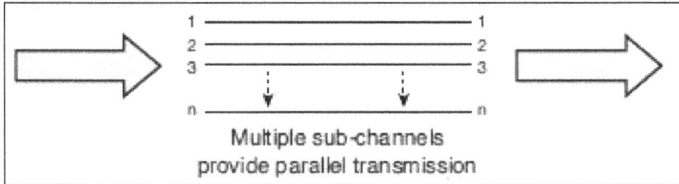

Multiple sub-channels
provide parallel transmission

OFDM send multitudes of data simultaneously in parallel.

## Radio Waves in Cellular Communication

In the most basic form, a cell phone is essentially a two-way radio, consisting of a radio transmitter and a radio receiver. When you chat with your friend on your cell phone, your phone converts your voice into an electrical signal, which is then transmitted via radio waves to the nearest cell tower. The network of cell towers then relays the radio wave to your friend's cell phone, which converts it to an electrical signal and then back to sound again. In the basic form, a cell phone works just like a walkie-talkie.

In additional to the basic function of voice calls, most modern cell phones come with additional functions such as web surfing, taking pictures, playing games, sending text messages and playing music. More sophisticated smart phones can perform similar functions of a portable computer.

## Radio Waves

Cell phones use radio waves to communicate. Radio waves transport digitized voice or data in the form of oscillating electric and magnetic fields, called the electromagnetic field (EMF). The rate of oscillation is called frequency. Radio waves carry the information and travel in air at the speed of light.

Cell phones transmit radio waves in all directions. The waves can be absorbed and reflected by surrounding objects before they reach the nearest cell tower. For example, when the phone is placed next to your head during a call, a significant portion (over half in many cases) of the emitted energy is absorbed into your head and body. In this event, much of the cell phone's EMF energy is wasted and no longer available for communication.

## Antenna

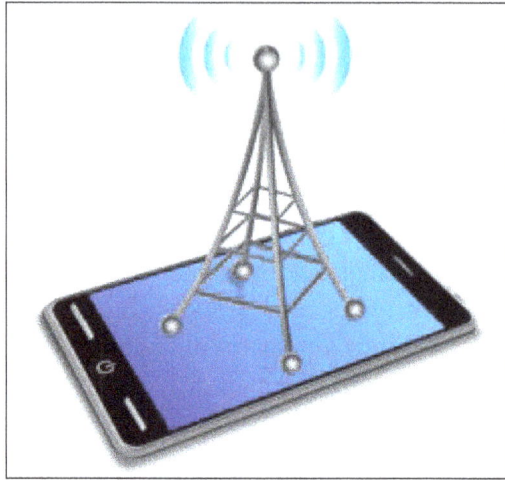

Cell phones contain at least one radio antenna in order to transmit or receive radio signals. An antenna converts an electric signal to the radio wave (transmitter) and vice versa (receiver). Some cell phones use one antenna as the transmitter and receiver while others, such as the iPhone 5, have multiple transmitting or receiving antennas.

An antenna is a metallic element (such as copper) engineered to be a specific size and shape for transmitting and receiving specific frequencies of radio waves. While older generation cell phones have external or extractable antennas, modern cell phones contain more compact antennas inside the device thanks to advanced antenna technologies. It's important to understand that any metallic components in the device (such as the circuit board and the metal frame for the iPhone) can interact with the transmission antenna(s) and contribute to the pattern of the transmitted signal.

Many modern smart phones also contain more than one type of antenna. In addition to the cellular antenna, they may also have Wi-Fi, Bluetooth and/or GPS antennas.

## Connectivity

A cell phone is a two-way wireless communication device and needs both the inbound signal (reception) and the outbound signal (transmission) to work. The magnitude of the received signal from the cell tower is called the "signal strength", which is commonly indicated by the "bars" on your phone. The connectivity between a cell phone and its cellular network depends on both signals and is affected by many factors, such as the distance between the phone and the nearest cell tower, the number of impediments between them and the wireless technology (e.g. GSM vs. CDMA). A poor reception (fewer bars) normally indicates a long distance and/or much signal interruption between the cell phone and the cell tower.

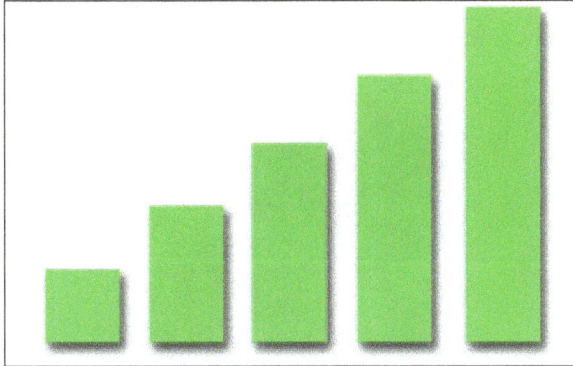

In order to conserve battery life, a cell phone will vary the strength of its transmitted signal and use only the minimum necessary to communicate with the nearest cell tower. When your cell phone has poor connectivity, it transmits a stronger signal in order to connect to the tower, and as a result your battery drains faster. That's why good connectivity not only reduces dropped calls, but also saves battery life.

## Microwaves

Microwaves are a form of electromagnetic radiation with wavelengths ranging from about one meter to one millimeter; with frequencies between 300 MHz (1 m) and 300 GHz (1 mm). Different sources define different frequency ranges as microwaves; the above broad definition includes both UHF and EHF (millimeter wave) bands. A more common definition in radio-frequency engineering is the range between 1 and 100 GHz (wavelengths between 0.3 m and 3 mm). In all cases, microwaves include the entire SHF band (3 to 30 GHz, or 10 to 1 cm) at minimum. Frequencies in the microwave range are often referred to by their IEEE radar band designations: S, C, X, Ku, K, or Ka band, or by similar NATO or EU designations.

The prefix *micro-* in *microwave* is not meant to suggest a wavelength in the micrometer range. Rather, it indicates that microwaves are "small" (having shorter wavelengths), compared to the radio waves used prior to microwave technology. The boundaries between far infrared, terahertz radiation, microwaves, and ultra-high-frequency radio waves are fairly arbitrary and are used variously between different fields of study.

Microwaves travel by line-of-sight; unlike lower frequency radio waves they do not diffract around hills, follow the earth's surface as ground waves, or reflect from the ionosphere, so terrestrial microwave communication links are limited by the visual horizon to about 40 miles (64 km). At the high end of the band they are absorbed by gases in the atmosphere, limiting practical communication distances to around a kilometer. Microwaves are widely used in modern technology, for example in point-to-point communication links, wireless networks, microwave radio relay networks, radar, satellite and spacecraft communication, medical diathermy and cancer treatment, remote sensing, radio astronomy, particle accelerators, spectroscopy, industrial heating, collision avoidance systems, garage door openers and keyless entry systems, and for cooking food in microwave ovens.

A telecommunications tower with a variety of dish antennas for microwave relay links. The apertures of the dishes are covered by plastic sheets (radomes) to keep out moisture.

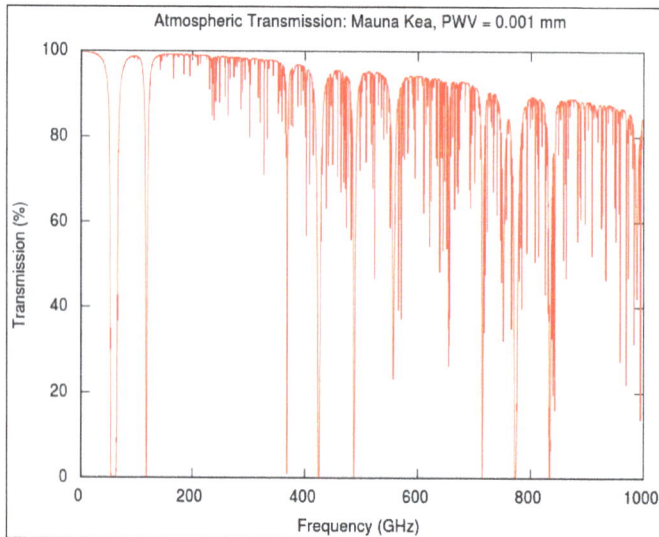

The atmospheric attenuation of microwaves and far infrared radiation in dry air with a precipitable water vapor level of 0.001 mm. The downward spikes in the graph correspond to frequencies at which microwaves are absorbed more strongly. This graph includes a range of frequencies from 0 to 1 THz; the microwaves are the subset in the range between 0.3 and 300 gigahertz.

## Electromagnetic Spectrum

Microwaves occupy a place in the electromagnetic spectrum with frequency above ordinary radio waves, and below infrared light:

| Electromagnetic spectrum | | | |
|---|---|---|---|
| Name | Wavelength | Frequency (Hz) | Photon energy (eV) |
| Gamma ray | < 0.02 nm | > 15 EHz | > 62.1 keV |
| X-ray | 0.01 nm – 10 nm | 30 EHz – 30 PHz | 124 keV – 124 eV |
| Ultraviolet | 10 nm – 400 nm | 30 PHz – 750 THz | 124 eV – 3 eV |
| Visible light | 390 nm – 750 nm | 770 THz – 400 THz | 3.2 eV – 1.7 eV |
| Infrared | 750 nm – 1 mm | 400 THz – 300 GHz | 1.7 eV – 1.24 meV |
| Microwave | 1 mm – 1 m | 300 GHz – 300 MHz | 1.24 meV – 1.24 µeV |
| Radio | 1 m – 100 km | 300 MHz – 3 kHz | 1.24 µeV – 12.4 feV |

In descriptions of the electromagnetic spectrum, some sources classify microwaves as radio waves, a subset of the radio wave band; while others classify microwaves and radio waves as distinct types of radiation. This is an arbitrary distinction.

## Propagation

Microwaves travel solely by line-of-sight paths; unlike lower frequency radio waves, they do not travel as ground waves which follow the contour of the Earth, or reflect off the ionosphere (skywaves). Although at the low end of the band they can pass through building walls enough for useful reception, usually rights of way cleared to the first Fresnel zone are required. Therefore, on the surface of the Earth, microwave communication links are limited by the visual horizon to about 30–40 miles (48–64 km). Microwaves are absorbed by moisture in the atmosphere, and the attenuation increases with frequency, becoming a significant factor (rain fade) at the high end of the band. Beginning at about 40 GHz, atmospheric gases also begin to absorb microwaves, so above this frequency microwave transmission is limited to a few kilometers. A spectral band structure causes absorption peaks at specific frequencies. Above 100 GHz, the absorption of electromagnetic radiation by Earth's atmosphere is so great that it is in effect opaque, until the atmosphere becomes transparent again in the so-called infrared and optical window frequency ranges.

## Troposcatter

In a microwave beam directed at an angle into the sky, a small amount of the power will be randomly scattered as the beam passes through the troposphere. A sensitive receiver beyond the horizon with a high gain antenna focused on that area of the troposphere can pick up the signal. This technique has been used at frequencies between 0.45 and 5 GHz in tropospheric scatter (troposcatter) communication systems to communicate beyond the horizon, at distances up to 300 km.

## Antennas

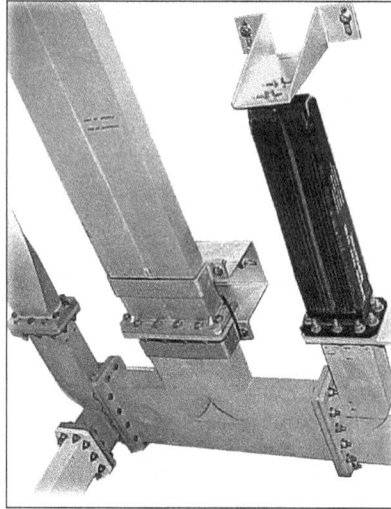

Waveguide is used to carry microwaves. Example of waveguides and a diplexer in an air traffic control radar.

The short wavelengths of microwaves allow omnidirectional antennas for portable devices to be made very small, from 1 to 20 centimeters long, so microwave frequencies are widely used for wireless devices such as cell phones, cordless phones, and wireless LANs (Wi-Fi) access for laptops, and Bluetooth earphones. Antennas used include short whip antennas, rubber ducky antennas, sleeve dipoles, patch antennas, and increasingly the printed circuit inverted F antenna (PIFA) used in cell phones.

Their short wavelength also allows narrow beams of microwaves to be produced by conveniently small high gain antennas from a half meter to 5 meters in diameter. Therefore, beams of microwaves are used for point-to-point communication links, and for radar. An advantage of narrow beams is that they do not interfere with nearby equipment using the same frequency, allowing frequency reuse by nearby transmitters. Parabolic ("dish") antennas are the most widely used directive antennas at microwave frequencies, but horn antennas, slot antennas and dielectric lens antennas are also used. Flat microstrip antennas are being increasingly used in consumer devices. Another directive antenna practical at microwave frequencies is the phased array, a computer-controlled array of antennas which produces a beam which can be electronically steered in different directions.

At microwave frequencies, the transmission lines which are used to carry lower frequency radio waves to and from antennas, such as coaxial cable and parallel wire lines, have excessive power losses, so when low attenuation is required microwaves are carried by metal pipes called waveguides. Due to the high cost and maintenance requirements of waveguide runs, in many microwave antennas the output stage of the transmitter or the RF front end of the receiver is located at the antenna.

## Design and Analysis

The term *microwave* also has a more technical meaning in electromagnetics and circuit theory. Apparatus and techniques may be described qualitatively as "microwave" when the wavelengths of signals are roughly the same as the dimensions of the circuit, so that lumped-element circuit theory is inaccurate, and instead distributed circuit elements and transmission-line theory are more useful methods for design and analysis.

As a consequence, practical microwave circuits tend to move away from the discrete resistors, capacitors, and inductors used with lower-frequency radio waves. Open-wire and coaxial transmission lines used at lower frequencies are replaced by waveguides and stripline, and lumped-element tuned circuits are replaced by cavity resonators or resonant stubs. In turn, at even higher frequencies, where the wavelength of the electromagnetic waves becomes small in comparison to the size of the structures used to process them, microwave techniques become inadequate, and the methods of optics are used.

## Microwave Sources

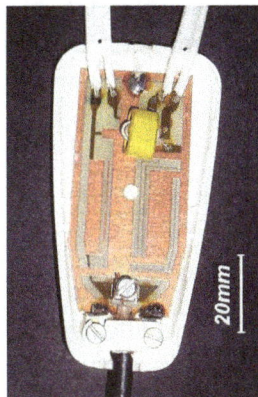

Cutaway view inside a cavity magnetron as used in a microwave oven *(left)*. Antenna splitter: microstrip techniques become increasingly necessary at higher frequencies *(right)*.

Disassembled radar speed gun. The grey assembly attached to the end of the copper-colored horn antenna is the Gunn diode which generates the microwaves.

High-power microwave sources use specialized vacuum tubes to generate microwaves. These devices operate on different principles from low-frequency vacuum tubes, using the ballistic motion of electrons in a vacuum under the influence of controlling electric or magnetic fields, and include the magnetron (used in microwave ovens), klystron, traveling-wave tube (TWT), and gyrotron. These devices work in the density modulated mode, rather than the current modulated mode. This means that they work on the basis of clumps of electrons flying ballistically through them, rather than using a continuous stream of electrons.

Low-power microwave sources use solid-state devices such as the field-effect transistor (at least at lower frequencies), tunnel diodes, Gunn diodes, and IMPATT diodes. Low-power sources are available as benchtop instruments, rackmount instruments, embeddable modules and in card-level formats. A maser is a solid state device which amplifies microwaves using similar principles to the laser, which amplifies higher frequency light waves.

All warm objects emit low level microwave black-body radiation, depending on their temperature, so in meteorology and remote sensing microwave radiometers are used to measure the temperature of objects or terrain. The sun and other astronomical radio sources such as Cassiopeia A emit low level microwave radiation which carries information about their makeup, which is studied by radio astronomers using receivers called radio telescopes. The cosmic microwave background radiation (CMBR), for example, is a weak microwave noise filling empty space which is a major source of information on cosmology's Big Bang theory of the origin of the Universe.

## Microwave Frequency Bands

Bands of frequencies in the microwave spectrum are designated by letters. Unfortunately, there are several incompatible band designation systems, and even within a system the frequency ranges corresponding to some of the letters vary somewhat between different application fields. The letter system had its origin in World War 2 in a top secret U.S. classification of bands used in radar sets; this is the origin of the oldest letter system, the IEEE radar bands. One set of microwave frequency bands designations by the Radio Society of Great Britain (RSGB), is tabulated below:

| Microwave Frequency Bands | | | |
|---|---|---|---|
| Designation | Frequency Range | Wavelength Range | Typical Uses |
| L band | 1 to 2 GHz | 15 cm to 30 cm | Military telemetry, GPS, mobile phones (GSM), amateur radio. |
| S band | 2 to 4 GHz | 7.5 cm to 15 cm | Weather radar, surface ship radar, and some communications satellites (microwave ovens, microwave devices/communications, radio astronomy, mobile phones, wireless LAN, Bluetooth, zigbee, GPS, amateur radio). |
| C band | 4 to 8 GHz | 3.75 cm to 7.5 cm | Long-distance radio telecommunications. |
| X band | 8 to 12 GHz | 25 mm to 37.5 mm | Satellite communications, radar, terrestrial broadband, space communications, amateur radio, molecular rotational spectroscopy. |
| $K_u$ band | 12 to 18 GHz | 16.7 mm to 25 mm | Satellite communications, molecular rotational spectroscopy. |
| K band | 18 to 26.5 GHz | 11.3 mm to 16.7 mm | Radar, satellite communications, astronomical observations, automotive radar, molecular rotational spectroscopy. |
| $K_a$ band | 26.5 to 40 GHz | 5.0 mm to 11.3 mm | Satellite communications, molecular rotational spectroscopy. |
| Q band | 33 to 50 GHz | 6.0 mm to 9.0 mm | Satellite communications, terrestrial microwave communications, radio astronomy, automotive radar, molecular rotational spectroscopy. |
| U band | 40 to 60 GHz | 5.0 mm to 7.5 mm | |
| V band | 50 to 75 GHz | 4.0 mm to 6.0 mm | Millimeter wave radar research, molecular rotational spectroscopy and other kinds of scientific research. |
| W band | 75 to 110 GHz | 2.7 mm to 4.0 mm | Satellite communications, millimeter-wave radar research, military radar targeting and tracking applications, and some non-military applications, automotive radar. |
| F band | 90 to 140 GHz | 2.1 mm to 3.3 mm | SHF transmissions: Radio astronomy, microwave devices/communications, wireless LAN, most modern radars, communications satellites, satellite television broadcasting, DBS, amateur radio. |
| D band | 110 to 170 GHz | 1.8 mm to 2.7 mm | EHF transmissions: Radio astronomy, high-frequency microwave radio relay, microwave remote sensing, amateur radio, directed-energy weapon, millimeter wave scanner. |

The term P band is sometimes used for UHF frequencies below the L band but is now obsolete per IEEE Std 521.

When radars were first developed at K band during World War II, it was not known that there was a nearby absorption band (due to water vapor and oxygen in the atmosphere). To avoid this problem, the original K band was split into a lower band, $K_u$, and upper band, $K_a$.

## Microwave Frequency Measurement

Absorption wavemeter for measuring in the $K_u$ band.

Microwave frequency can be measured by either electronic or mechanical techniques.

Frequency counters or high frequency heterodyne systems can be used. Here the unknown frequency is compared with harmonics of a known lower frequency by use of a low frequency generator, a harmonic generator and a mixer. Accuracy of the measurement is limited by the accuracy and stability of the reference source.

Mechanical methods require a tunable resonator such as an absorption wavemeter, which has a known relation between a physical dimension and frequency.

In a laboratory setting, Lecher lines can be used to directly measure the wavelength on a transmission line made of parallel wires, the frequency can then be calculated. A similar technique is to use a slotted waveguide or slotted coaxial line to directly measure the wavelength. These devices consist of a probe introduced into the line through a longitudinal slot, so that the probe is free to travel up and down the line. Slotted lines are primarily intended for measurement of the voltage standing wave ratio on the line. However, provided a standing wave is present, they may also be used to measure the distance between the nodes, which is equal to half the wavelength. Precision of this method is limited by the determination of the nodal locations.

## Microwaves in Cellular Communication

Past decades have seen the explosive growth of wireless communications. A sequence of breakthroughs such as MIMO, capacity achieving codes, millimeter-wave communications and small-cell networks have achieved gigabit speeds for wireless access. As wireless and wire access speeds are becoming comparable, mobile devices, including smartphones and tablet and laptop computers, have replaced desktop computers as the dominant platforms for Internet access. In contrast, the advancements in battery technologies have been much slower. The resultant short battery lives require mobile devices to be periodically tethered to the grid for battery recharging. The cables for recharging are the last barrier for the devices to attain true mobility and thus called "last wires". As mobile services have penetrated different fields of the modern society such as banking, health care and civil defense, the interruption due to dead batteries can cause issues far more severe than merely inconvenience, such as financial loss and threats to health and public safety. Moreover, the production of billions of non-recyclable chargers per year poses a serious environmental issue. The urgency of addressing these issues and the existence of many market opportunities have recently motivated both the industry and academia to direct huge effort and funding towards developing technologies for wireless power transfer. Breakthroughs in such technologies will solve the grand ICT challenge of cutting the "last wires".

The idea of wireless power transfer using radio waves was first conceived and experimented by Nicola Tesla in 1899. However, the area did not pick up till 1960s when microwave technologies rapidly advanced, opening an active research field called microwave power transfer (MPT). In particular, the availability of large-scale antenna arrays and high-power microwave generators enables beaming of high power in a desirable direction. Moreover, the invention of rectifying antennas (rectennas) renders energy-conversion loss to a level practically negligible. Many advanced MPT systems have been designed such as wirelessly powered airborne vehicles that require no refueling and solar power satellites. However, the enormous antenna arrays (e.g., arrays with diameters of hundreds of meters) that are instrumental for efficient MPT in such systems are impractical for everyday-life applications. This issue together with safety concerns has been delaying the commercialization of MPT. On the contrary, without such issues, non-radiative technologies for wireless power transfer, namely inductive coupling and resonant coupling between two coils, have been standardized and widely commercialized in mobile devices, home appliances and electric vehicles. However, such technologies have the drawbacks of extremely short transfer distances (typically no more than a meter) and lack of support for mobility.

With long propagation ranges and support of mobility and multicasting, MPT appears to be a promising candidate technology for cutting the "last wires" if the two main challenges, high propagation loss and safety concerns, can be overcome. Both relying on microwaves as transmission vehicles, MPT and wireless communication have interwound R&D histories, yielding many common techniques and theories e.g., beamforming and propagation. The similarity allows communication techniques and network designs to be applied to tackle the mentioned MPT challenges. Specifically, sophisticated signal processing techniques for channel estimation, power control and adaptive beamforming can be adopted to ensure safety in MPT. Moreover, the latest breakthroughs in wireless communications, namely small cells, transmission using large-scale antenna arrays and millimeter-wave communications will dramatically reduce transmission distances and enable sharp beamforming, which will suppress propagation loss and achieve high powertransfer efficiencies. The availability of enabling technologies suggests that the time has come for cutting the "last wires", opening up the new area of wirelessly powered communications (WPC) with many exciting new research opportunities and applications. This topic will introduce WPC by discussing its key features, answering a set of frequently asked questions, and identifying the key design challenges.

## Key Features of Wirelessly Powered Communications

### Power Beamforming

Efficient MPT hinges on concentrating radiated power in the direction of a target mobile by forming a microwave beam. Such beamforming for the sole purpose of power transfer is called power beamforming. A beam can be formed using an antenna array (or aperture antenna). The elements of the array are arranged with separation no larger than a half wavelength so as to avoid "grating lobes" (multiple beams). Under this constraint, the beam sharpness increases with the array size (or equivalently the number of elements). Sharp beamforming and short propagation distances are the key conditions for efficient MPT. They can be achieved by two corresponding latest wireless-communication technologies, namely large-scale antenna arrays (with hundreds to thousands of antenna elements) and small cells, currently under extensive development and expected to be deployed in next-generation wireless networks.

With the array size fixed, the beam sharpness can be increased by scaling up the carrier frequency and correspondingly packing more antennas into the array. Traditional MPT without a dedicated spectrum uses the carrier frequency of either 2.4 GHz or 5.8 GHz in the ISM band. However, with the rapid advancement of millimeter-wave communication, the MPT embedded in WPC can be operated in the 60- GHz bandwidth in the near future. Such high frequencies enable ultra-sharp beamforming even when the array size is small, leading to dramatically improved power-transfer efficiencies.

## Power-transfer Channel and Beam Efficiency

Rich scattering is typical in a wireless communication channel and can be combined with transmit and receive antenna arrays to support multiple parallel data streams without requiring additional bandwidth. In contrast, free-space propagation is essential for power beamforming, since a scatterer can disperse power beam and cause the transfer efficiency to drop dramatically. Thus, a power-transfer channel refers to one over free space. The propagation distance ranges from the near field, where the distance is comparable with the transmit array dimension, to the far field. The factors determining the propagation loss include: 1) the apertures of the transmit and receive arrays, denoted as $A_t$ and $A_r$, respectively, 2) the wavelength $\lambda$ and 3) the propagation distances $d$ as elaborated in the sequel.

The end-to-end MPT efficiency is equal to the product of three efficiencies: 1) DC-to-RF power-conversion efficiency, 2) beam efficiency defined as the ratio between the received and radiated powers, and 3) RF-toDC power-conversion efficiency. The state-of-the-art microwave generators and rectennas can achieve closeto-one values (e.g., 80%) for the first and third efficiencies, respectively. Therefore, the beam efficiency is the bottleneck for efficient MPT over long distances.

Consider a pair of transmit/receive circular aperture antennas (that can be replaced by arrays with the same apertures) facing each other over a power-transfer channel. For this scenario, the beam efficiency can be accurately approximated as:

$$\text{Beam Efficiency } = 1 - e^{-\beta}$$

where $\beta$ is given as:

$$\beta = \frac{A_t A_r}{(\lambda d)^2}.$$

Note that (2) is equivalent to the Friis equation for far-field transmission. The propagation as described in (1) covers both the near and far fields. For the far field where $\beta$ is small (large d), the propagation loss is approximately equal to $\beta$, thus following the Friis transmission equation. For the near field where d is small and hence $\beta$ is large, the beam efficiency is close to one. The beam efficiency is plotted in Fig. 1 as a function of the ratio between the transfer distance and the receiver antenna radius, where the carrier frequency is 2.4 GHz. Next, (1) suggests the trade-offs between the MPT parameters $A_r$, $A_t$, $\lambda$ and d. In particular, for a given beam efficiency, doubling the transmit-array radius or the carrier frequency doubles the transfer distance or supports recharging of smaller (half-size) mobiles. For instance, scaling up the frequency from 2.4 GHz to 60 GHz in the millimeter band increases the power-transfer distance by 25 times. This, however, requires the numbers of transmit/receive antennas to grow by the square of this factor if the aperture antennas are replaced by antenna arrays.

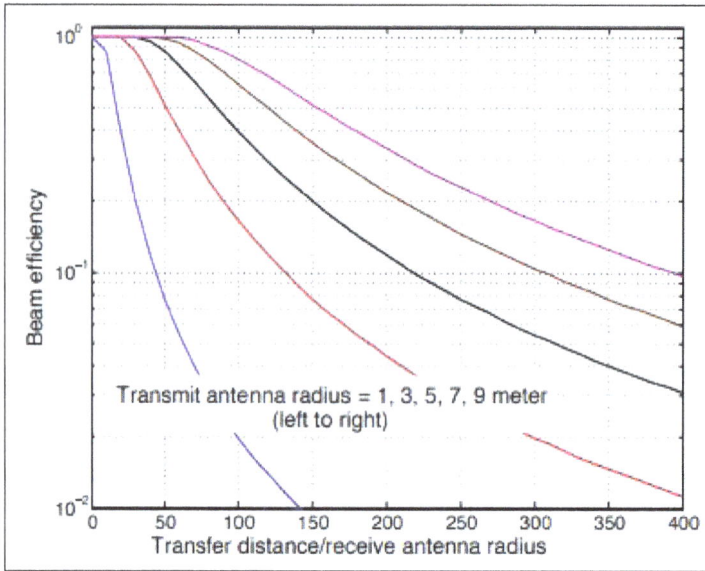

Beam efficiency for MPT versus the ratio between the transfer distance and the receiver antenna radius for a carrier frequency of 2.4 GHz.

## Mobile Architecture for WPC

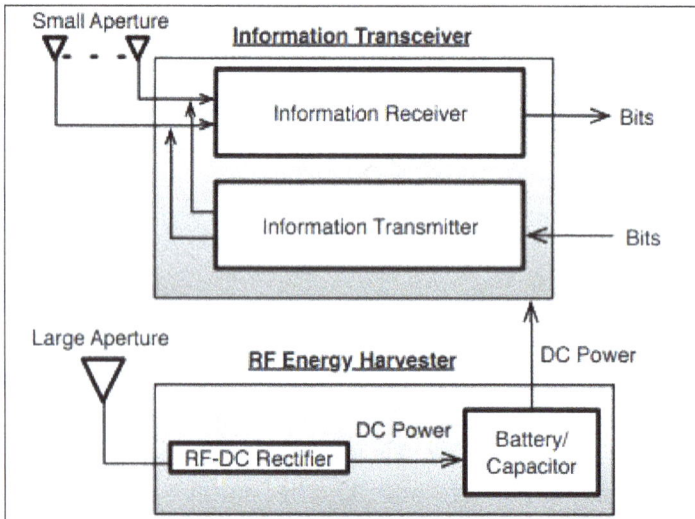

Mobile architecture for WPC.

A traditional mobile device comprises an information transceiver powered by a rechargeable battery. For WPC, a RF energy harvester is included in the mobile device for harvesting energy from the incident microwave signal to power the transceiver as shown in figure. The design of an energy harvester is rather simple and consists of a rectifying circuit for converting the RF signal at the antenna output to DC power that is stored using a rechargeable battery or a super capacitor. The most popular and efficient design of the RF energy harvester uses a rectenna which integrates a single antenna and a rectifying circuit.

The functionality of the antennas used by the information transceiver and energy harvester is different. The array attached to the transceiver enables multi-antenna communication and array processing (e.g., receive beamforming and interference nulling). Therefore, it is desirable to have as many (small) antenna elements as possible. On the other hand, the rectenna requires to capture as much incident power as possible, hence, the rectenna design aims for the largest possible antenna aperture. This is an important tradeoff for designing a WPC receiver under a form-factor constraint.

## Simultaneous Wireless Information and Power Transfer

Since the information transceiver and energy harvester have separate antennas and circuits, the mobile architecture supports SWIPT. There exist three designs of SWIPT system, namely integrated SWIPT, closed-loop SWIPT, and decoupled SWIPT, which are described as follows. Integrated SWIPT is the simplest design where power and information are extracted by the mobile from the same modulated microwave transmitted by a base station (BS). For this design, information transfer (IT) and power transfer (PT) distances are constrained to be equal. Closed-loop SWIPT consists of downlink PT and uplink IT. The signal power received at the BS originates from the BS radiated power and its closed-loop propagation (downlink+uplink) incurs double attenuation. Thus closed-loop SWIPT only supports very short ranges and is unsuitable for cell-edge mobiles. Last, a decoupled-SWIPT system in figure (c) builds on the traditional communication system to include an additional special station, called power beacon (PB), dedicated for MPT to mobiles. PT and IT are orthogonalized by using different frequency bands or time slots to avoid interference, giving the name decoupled SWIPT. Unlike BSs, PBs require no backhaul links and the resultant low cost allows dense deployment of PBs to enable efficient MPT.

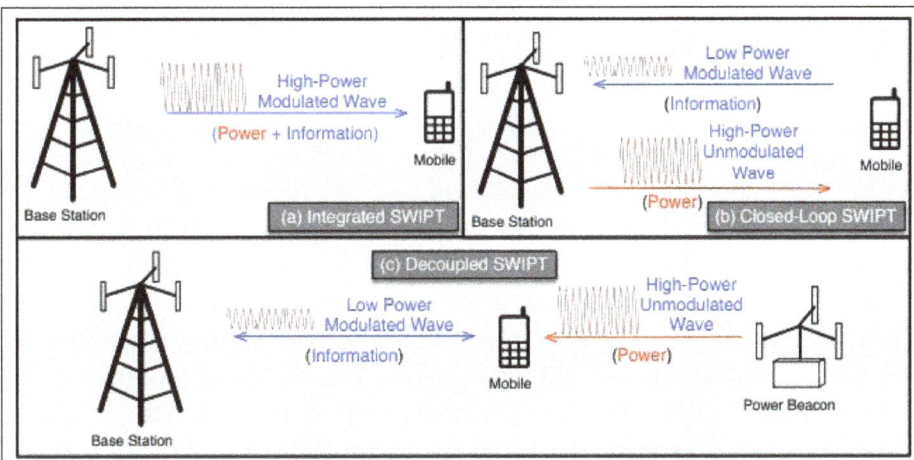

Three system configurations for SWIPT: (a) integrated SWIPT,
(b) closed-loop SWIPT and (c) decoupled SWIPT.

## Wirelessly Powered Communications: Designs and Challenges

### Pilot Signal Design for Retrodirective Beam Control

Power beamforming for MPT typically uses a phase array and the technique called retrodirective beam control that automatically steers a beam in the reverse direction of the incident pilot signal sent by the mobile by exploiting channel reciprocity. The simple procedure for retrodirective beam control is as follows:

- The receiver transmits a pilot signal.

- The transmitter computes the phase shift of the output of each antenna by comparing it with a local reference signal.

- The phase shift for each antenna is conjugated and applied to the phase shifter for transmission.

An important aspect of implementing retrodirective beam control for WPC is the design of pilot sequences. They should be designed to initiate MPT for multiple mobiles at the same time. However, mobiles in different cells may transmit non-orthogonal or identical pilot sequences, resulting in pilot contamination which lays a fundamental limit for the performance of a cellular network equipped with large-scale antenna arrays. In WPC networks, pilot contamination not only degrades the IT performance but also decreases PT efficiency. To be specific, receiving multiple pilot signals can cause the retrodirective beamformer for MPT to auto-reflect multiple beams towards both the intended and unintended mobiles, which reduces the beam efficiency to the former and, more importantly, causes safety threats to people in unintended directions. Tackling pilot contamination continues to be a key challenge in designing WPC networks.

In addition, the power of pilot signals and duty cycle of pilot transmission should be designed to optimize the tradeoffs between the PT efficiency, training overhead and mobile energy consumption.

### Safety Measures

Measures for ensuring safe PT is a unique and important aspect of designing WPC systems. The retrodirective beamformer has the safety feature that it can automatically dephase a beam when it is intercepted by an object such as a human body. This measure, however, does not protect people who are near a target mobile but do not intercept the beam. Thus, additional safety measures are required. For example, guard zones can be created around not only the PBs but also the mobiles using technologies such as microwave life detection. Other technologies such as surveillance cameras, radar tracking and network localization can be deployed for accurate human detection and thereby further enhancing the safety in WPC.

## Efficient and Safe Power Transfer using Multiple Coordinated Power Beacons

In a WPC network deploying dense PBs, a single mobile can be powered by multiple coordinated PBs based on the idea proposed in. The PBs surrounding the target mobile form multiple incoming power beams from omni-directions that are coherently combined at the mobile location due to beacon coordination. PT using coordinated beacons is safe for two reasons. First, many incoming beams from different directions enable the detection of human presence at practically any arbitrary location near the target mobile, thereby overcoming the drawback of stand-alone PBs. Next, incident power beams from omni-directions have the combined effect of concentrating transmitted power at the mobile and very low power density at other locations. Moreover, multibeams improves the chance of finding lines-of-sight for efficient MPT. As the combined result, MPT using coordinated beacons also improves the PT efficiency.

## WPC Network Architecture

A WPC network is designed to deliver two types of services, high-speed wireless access and MPT, to mobile devices and its performance is measured by the coverages of both services. As illustrated in figure, the WPC network comprises BSs, PBs and mobile devices. The main role of BSs is to provide network-wise wirelessaccess coverage while they can also support SWIPT to nearby mobiles. Full MPT coverage is achieved by deploying dense PBs for supporting MPT to mobiles. Mobiles can be separated into receiving and transmitting mobiles. It is more challenging to wirelessly power transmitting mobiles since they require additional power for transmission while both types of mobiles consume circuit power.

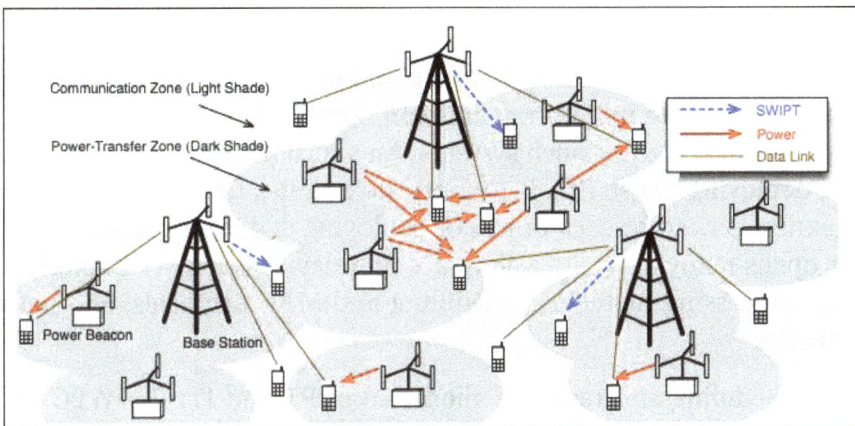

WPC network architecture.

For designing a WPC network, one of the first challenges is to understand the required densities of BSs and PBs for providing network coverage for both wireless access and wireless power. Recently, the tradeoff between these densities was quantified in by modeling the WPC network using stochastic geometry and under reliability constraints on the network services. This tractable approach can be extended to design WPC networks with more complex architectures such as heterogeneous BSs/PBs. Apart from the fixed deployment of PBs, mobile PBs can be also deployed to support wider coverage with fewer PBs or more efficient MPT by shortening the transfer distances. One challenge there is to design the optimal routing for each mobile PB.

## WPC Protocols and Techniques

Compared with traditional wireless networks, the addition of PBs and the interaction between IT and PT enrich the WPC network architecture and the operational modes of network nodes. As a result, traditional communication protocols and techniques must be thoroughly redesigned to enable efficient WPC. Several key challenges are identified as follows, which point to promising research directions.

- Cognitive WPC: The principle of cognitive radios can be applied to design cognitive WPC systems to enable seamless integration of PT and IT and accommodate passive secondary nodes. In particular, a cognitive PB can sense the spectrum and choose a proper subset of frequency sub-channels for MPT to avoid interfering with IT and at the same time reduce powerspectrum density to meet the safety requirements set by authorities.

- Cooperative PB/BS clustering: Grouping PBs/BSs for cooperation enhances the PT efficiency besides mitigates interference in IT. However, PB/BS clustering is much more complex than that for traditional multi-cell cooperation due to many new factors for consideration including multi-user beam efficiencies, BS modes (SWIPT or IT only) and wireless signaling overhead between backhaul-less PBs.

- Relay Assisted WPC: In WPC systems with relatively sparse BSs/PBs, mobiles far away from them receive double penalties: lower PT efficiencies but larger power required for uplink transmission. Thus, it is critical to address the issue of fairness in designing such systems. An alternative cost-effect approach apart from deploying dense PBs is to motivate mobiles to cooperate by relaying information/power for each others or deploying dedicated passive relay stations. This opens many new research issues on relay-assisted WPC ranging from the signal processing methods, scheduling and MAC protocols, and network performance.

- Joint scheduling and resource allocation of PT and IT: For WPC networks of low-complexity and low-power devices, the communication protocols are often simple and predetermined. In contrast, for scenarios where the mobiles are able to handle complex algorithms, the optimal solution is to design and deploy intelligent communication and resource allocation algorithms for mobiles, PBs and BSs that are adapted to the dynamic states of mobile energy storage, data queues, channels and beam efficiencies.

## Towards Truly Mobile Communications

Cutting the "last wires" of mobile devices will endow them the long desired immortality, which will bring users convenience, strengthen the reliability of widespread mobile services and create a huge range of market opportunities. This task is far more than straightforward implementation of the MPT technology but requires a seamless integration between information and power transfers. As a result, many new research challenges arise including designing network architectures for enabling SWIPT, achieving highly efficient and safe MPT to mobile devices, and revamping traditional communication techniques, such as cooperation, cognitive radios and adaptive transceivers, to integrate power transfer into communication networks. This leads to a newly emerged area called wirelessly powered communications. It is through the advancements in this area and relevant areas such as energy scavenging, batteries and low-power electronics that the tens of billions of devices to be deployed in the coming decade will be free from the "last wires" and attain true mobility.

## Infrared Waves

Infrared (IR) is a wireless mobile technology used for device communication over short ranges. IR communication has major limitations because it requires line-of-sight, has a short transmission range and is unable to penetrate walls. IR transceivers are quite cheap and serve as short-range communication solutions.

Infrared transmission refers to energy in the region of the electromagnetic radiation spectrum at wavelengths longer than those of visible light, but shorter than those of radio waves. Correspondingly, infrared frequencies are higher than those of microwaves, but lower than those of visible light.

Scientists divide the infrared radiation (IR) spectrum into three regions. The wavelengths are specified in microns (symbolized $\mu$, where $1 \mu = 10^{-6}$ meter) or in nanometers (abbreviated nm, where $1$ nm $= 10^{-9}$ meter $= 0.001$ $5$). The *near IR band* contains energy in the range of wavelengths closest to the visible, from approximately $0.750$ to $1.300$ $5$ ($750$ to $1300$ nm). The *intermediate IR band* (also called the *middle IR band*) consists of energy in the range $1.300$ to $3.000$ $5$ ($1300$ to $3000$ nm). The *far IR band* extends from $2.000$ to $14.000$ $5$ ($3000$ nm to $1.4000$ x $10^4$ nm).

Infrared is used in a variety of wireless communications, monitoring, and control applications. Here are some examples:

- Home-entertainment remote-control boxes.

- Wireless (local area networks).

- Links between notebook computers and desktop computers.

- Cordless modem.

- Intrusion detectors.

- Motion detectors.

- Fire sensors.

- Night-vision systems.

- Medical diagnostic equipment.

- Missile guidance systems.

- Geological monitoring devices.

Transmitting IR data from one device to another is sometimes referred to as beaming.

## Infrared Waves in Cellular Communication

Infrared band of the electromagnet corresponds to 430THz to 300GHz and a wavelength of 980nm. The propagation of light waves in this band can be used for a communication system (for transmission and reception) of data. This communication can be between two portable devices or between a portable device and a fixed device.

There are two types of Infrared communication:

- Point to Point: It requires a line of sight between the transmitter and a receiver. In other words the transmitter and the receiver should be pointed to each other and there shouldn't be any obstacles between them. Example is the remote control communication.

- Diffuse Point: It doesn't require any line of sight and the link between the transmitter and the receiver is maintained by reflecting or bouncing of the transmitted signal by surfaces like ceilings, roof, etc. Example is the wireless LAN communication system.

## Advantages of IR Communication

- Security: Infrared communication has high directionality and can identify the source as different sources emit radiation of different frequencies and thus the risk of information being diffused is eliminated.

- Safety: Infrared radiation is not harmful to human beings. Hence infrared communication can be used at any place.

- High Speed data Communication: The data rate of Infrared communication is about 1Gbps and can be used for sending information like video signal.

## IR Communication Basics

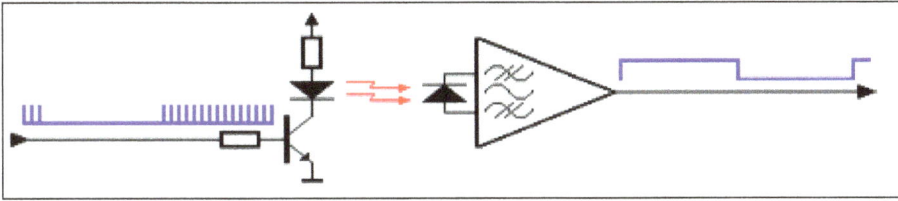

IR communication principle.

## IR Transmission

The transmitter of an IR LED inside its circuit, which emits infrared light for every electric pulse given to it. This pulse is generated as a button on the remote is pressed, thus completing the circuit, providing bias to the LED.

The LED on being biased emits light of the wavelength of 940nm as a series of pulses, corresponding to the button pressed. However since along with the IR LED many other sources of infrared light such as us human beings, light bulbs, sun, etc, the transmitted information can be interfered. A solution to this problem is by modulation. The transmitted signal is modulated using a carrier frequency of 38 KHz (or any other frequency between 36 to 46 KHz). The IR LED is made to oscillate at this frequency for the time duration of the pulse. The information or the light signals are pulse width modulated and are contained in the 38 KHz frequency.

## IR Reception

The receiver consists of a photodetector which develops an output electrical signal as light is incident on it. The output of the detector is filtered using a narrow band filter that discards all the frequencies below or above the carrier frequency (38 KHz in this case). The filtered output is then given to the suitable device like a Microcontroller or a Microprocessor which controls devices like a PC or a Robot. The output from the filters can also be connected to the Oscilloscope to read the pulses.

## Parts of IR Communication System

### IR Transmittor- IR Sensor

The sensors could be utilized as a part of measuring the radiation temperature without any contact. For different radiation temperature ranges various filters are available. An infrared (IR) sensor is an electronic device that radiates or locates infrared radiation to sense some part of its surroundings. They are undetectable to human eyes.

An infrared sensor could be considered a Polaroid that briefly recalls how an area's infrared radiation shows up. It is very regular for an infrared sensor to be coordinated into movement indicators like those utilized as a feature of private or business security systems. An IR sensor is shown in figure basically it has two terminals positive and negative. These sensors are undetectable to human eyes. They can measure the heat of an object and also identify movement. The region wavelength roughly from 0.75μm to 1000 μm is the IR region. The wavelength region of 0.75μm to 3 μm is called close infrared, the region from 3 μm to 6 μm is called mid infrared and the region higher than 6 μm is called far infrared. IR sensors emits at a frequency of 38 KHz.

IR Sensor.

## Features of IR Sensor

- Input voltage: 5VDC.

- Sensing Range: 5cm.

- Output signal: Analog voltage.

- Emitting element: Infrared LED.

## Example Interfacing Circuit of IR Diode and Photodiode

IR sensors mostly used in radiation thermometer, gas analyzers, industrial applications, IR imaging devices, tracking, and human body detection, communication and health hazards.

## Here is a Brief Description of IR and Photo Diode Sensing Switch

An IR diode is connected through a resistance to the dc supply. A photo diode is connected in reverse biased condition through a potential divider of a 10k variable resistance and 1k in series to the base of the transistor. While the IR rays fall on the reverse biased photo diode it conducts that causes a voltage at the base of the transistor.

The transistor then works like a switch while the collector goes to ground. Once the IR rays are obstructed the driving voltage is not available to the transistor thus its collector goes high. This low to high logic can be used for the microcontroller input for any action as per the program.

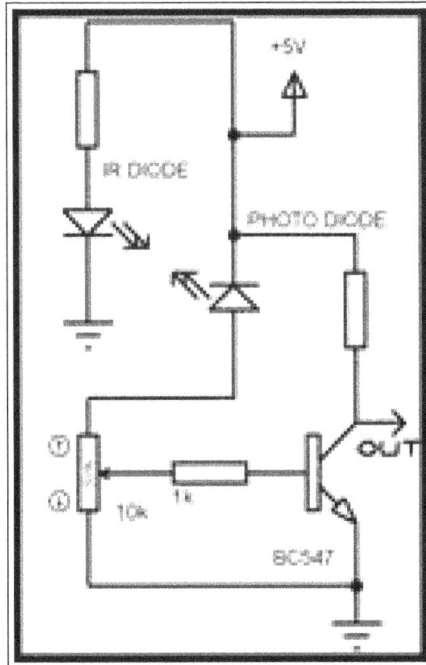

Circuit ir sensor.

## TSOP Sensor

TSOP is the standard IR remote control receiver series, supporting all major transmission codes. This is capable of receiving infrared radiation modulated at 38 kHz. IR sensors we have seen up to now working just for little short distance up to 6 cm. TSOP is sensitive to a specific frequency so its range is better contrast with ordinary photo diode. We can alter it up to 15 cm.

TSOP acts like as a receiver. It has three pins GND, Vs and OUT. GND is connected to common ground, Vs is connected to +5volts and OUT is connected to output pin. TSOP sensor has an inbuilt control circuit for amplifying the coded pulses from the IR transmitter. These are commonly used in TV remote receivers.

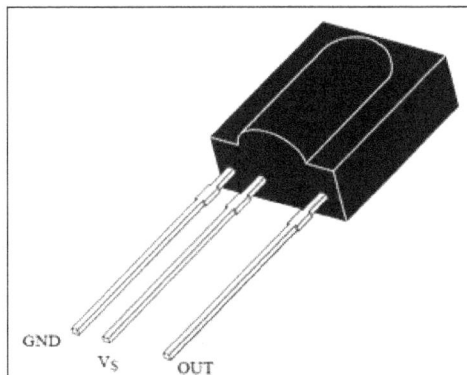

TSOP Sensor.

## Features

- The preamplifier and photo detector both are in single package.
- Internal filter for PCM frequency.
- Improved shielding against electrical field disturbance.
- TTL and CMOS compatibility.
- Output active low.
- Low power consumption.
- High immunity against ambient light.
- Continuous data transmission possible.

## Specifications

- Supply Voltage is −0.3-6.0 V.
- Supply Current is 5 mA.
- Output Voltage is −0.3-6.0 V.
- Output Current is 5 mA.
- Storage Temperature Range is −25-+85 °C.
- Operating Temperature Range is −25-+85°C.

The testing of TSOP is very simple. These are commonly used in TV remote receivers. TSOP consists of a PIN diode and pre-amplifier internally. Connect TSOP sensor as shown in circuit. A LED is connected through a resistance from the supply to output.

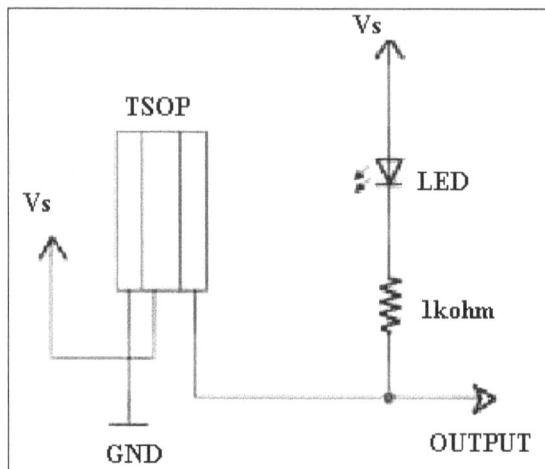

TSOP Sensor Circuit.

And then when we press the button of T.V. Remote control in front of the TSOP sensor, if LED starts blinking then our TSOP sensor and its connection is correct. The point when the output of TSOP is low i.e. at the time it appropriates IR signal from a source, with a centre frequency of 38 kHz, its output goes low.

TSOP sensor is used in our daily use TV, VCD, music system's remote control. Where IR rays are transmitted by pushing a button on remote which are received by TSOP receiver inside the equipment.

## Communication Satellites

A communications satellite is an artificial satellite that relays and amplifies radio telec ommunications signals via a transponder; it creates a communication channel between a source transmitter and a receiver at different locations on Earth. Communications satellites are used for television, telephone, radio, internet, and military applications. There are about 2,000 communications satellites in Earth's orbit, used by both private and government organizations. Many are in geostationary orbit 22,236 miles (35,785 km) above the equator, so that the satellite appears stationary at the same point in the sky, so the satellite dish antennas of ground stations can be aimed permanently at that spot and do not have to move to track it.

The high frequency radio waves used for telecommunications links travel by line of sight and so are obstructed by the curve of the Earth. The purpose of communications satellites is to relay the signal around the curve of the Earth allowing communication between widely separated geographical points. Communications satellites use a wide range of radio and microwave frequencies. To avoid signal interference, international or-ganizations have regulations for which frequency ranges or "bands" certain organizations are allowed to use. This allocation of bands minimizes the risk of signal interference.

### Satellite Orbit

Communications satellites usually have one of three primary types of orbit, while oth-er orbital classifications are used to further specify orbital details.

- Geostationary satellites have a *geostationary orbit* (GEO), which is 22,236 miles (35,785 km) from Earth's surface. This orbit has the special characteristic that the apparent position of the satellite in the sky when viewed by a ground observer does not change, the satellite appears to "stand still" in the sky. This is because the satellite's orbital period is the same as the rotation rate of the Earth. The advantage of this orbit is that ground antennas do not have to track the satellite across the sky, they can be fixed to point at the location in the sky the satellite appears.

- Medium Earth orbit (MEO) satellites are closer to Earth. Orbital altitudes range from 2,000 to 36,000 kilometres (1,200 to 22,400 mi) above Earth.

- The region below medium orbits is referred to as *low Earth orbit* (LEO), and is about 160 to 2,000 kilometres (99 to 1,243 mi) above Earth.

As satellites in MEO and LEO orbit the Earth faster, they do not remain visible in the sky to a fixed point on Earth continually like a geostationary satellite, but appear to a ground observer to cross the sky and "set" when they go behind the Earth. Therefore, to provide continuous communications capability with these lower orbits requires a larger number of satellites, so one will always be in the sky for transmission of communication signals. However, due to their relatively small distance to the Earth their signals are stronger.

## Low Earth Orbit

A low Earth orbit (LEO) typically is a circular orbit about 160 to 2,000 kilometres (99 to 1,243 mi) above the earth's surface and, correspondingly, a period (time to revolve around the earth) of about 90 minutes.

Because of their low altitude, these satellites are only visible from within a radius of roughly 1,000 kilometres (620 mi) from the sub-satellite point. In addition, satellites in low earth orbit change their position relative to the ground position quickly. So even for local applications, many satellites are needed if the mission requires uninterrupted connectivity.

Low-Earth-orbiting satellites are less expensive to launch into orbit than geostationary satellites and, due to proximity to the ground, do not require as high signal strength. (Recall that signal strength falls off as the square of the distance from the source, so the effect is dramatic.) Thus there is a tradeoff between the number of satellites and their cost.

In addition, there are important differences in the onboard and ground equipment needed to support the two types of missions.

## Satellite Constellation

A group of satellites working in concert is known as a satellite constellation. Two such constellations, intended to provide satellite phone services, primarily to remote areas, are the Iridium and Globalstar systems. The Iridium system has 66 satellites.

It is also possible to offer discontinuous coverage using a low-Earth-orbit satellite capable of storing data received while passing over one part of Earth and transmitting it later while passing over another part. This will be the case with the CASCADE system of Canada's CASSIOPE communications satellite. Another system using this store and forward method is Orbcomm.

## Medium Earth Orbit

A MEO is a satellite in orbit somewhere between 2,000 and 35,786 kilometres (1,243 and 22,236 mi) above the earth's surface. MEO satellites are similar to LEO satellites in functionality. MEO satellites are visible for much longer periods of time than LEO satellites, usually between 2 and 8 hours. MEO satellites have a larger coverage area than LEO satellites. A MEO satellite's longer duration of visibility and wider footprint means fewer satellites are needed in a MEO network than a LEO network. One disadvantage is that a MEO satellite's distance gives it a longer time delay and weaker signal than a LEO satellite, although these limitations are not as severe as those of a GEO satellite.

Like LEOs, these satellites do not maintain a stationary distance from the earth. This is in contrast to the geostationary orbit, where satellites are always 35,786 kilometres (22,236 mi) from the earth.

Typically the orbit of a medium earth orbit satellite is about 16,000 kilometres (10,000 mi) above earth. In various patterns, these satellites make the trip around earth in anywhere from 2 to 8 hours.

Example:

In 1962, the first communications satellite, Telstar, was launched. It was a medium earth orbit satellite designed to help facilitate high-speed telephone signals. Although it was the first practical way to transmit signals over the horizon, its major drawback was soon realized. Because its orbital period of about 2.5 hours did not match the Earth's rotational period of 24 hours, continuous coverage was impossible. It was apparent that multiple MEOs needed to be used in order to provide continuous coverage.

## Geostationary Orbit

To an observer on Earth, a satellite in a geostationary orbit appears motionless, in a fixed position in the sky. This is because it revolves around the Earth at Earth's own angular velocity (one revolution per sidereal day, in an equatorial orbit).

A geostationary orbit is useful for communications because ground antennas can be aimed at the satellite without their having to track the satellite's motion. This is relatively inexpensive.

In applications that require many ground antennas, such as DirecTV distribution, the savings in ground equipment can more than outweigh the cost and complexity of placing a satellite into orbit.

Examples:

- The first geostationary satellite was Syncom 3, launched on August 19, 1964, and used for communication across the Pacific starting with television coverage of the 1964 Summer Olympics. Shortly after Syncom 3, Intelsat I, aka *Early Bird*, was launched on April 6, 1965, and placed in orbit at 28° west longitude. It was the first geostationary satellite for telecommunications over the Atlantic Ocean.

- On November 9, 1966, Canada's first geostationary satellite serving the continent, Anik A1, was launched by Telesat Canada, with the United States following suit with the launch of Westar 1 by Western Union on April 13, 1967.

- On May 30, 1977, the first geostationary communications satellite in the world to be three-axis stabilized was launched: the experimental satellite ATS-6 built for NASA.

- After the launches of the Telstar through Westar 1 satellites, RCA Americom (later GE Americom, now SES) launched Satcom 1 in 1975. It was Satcom 1 that was instrumental in helping early cable TV channels such as WTBS (now TB S), HBO, CBN (now Freeform) and The Weather Channel become successful, because these channels distributed their programming to all of the local cable TV headends using the satellite. Additionally, it was the first satellite used by broadcast television networks in the United States, like ABC, NBC, and CBS, to distribute programming to their local affiliate stations. Satcom 1 was widely used because it had twice the communications capacity of the competing Westar 1 in America (24 transponders as opposed to the 12 of Westar 1), resulting in lower transponder-usage costs. Satellites in later decades tended to have even higher transponder numbers.

By 2000, Hughes Space and Communications (now Boeing Satellite Development Center) had built nearly 40 percent of the more than one hundred satellites in service worldwide. Other major satellite manufacturers include Space Systems/Loral, Orbital Sciences Corporation with the Star Bus series, Indian Space Research Organisation, Lockheed Martin (owns the former RCA Astro Electronics/GE Astro Space business), Northrop Grumman, Alcatel Space, now Thales Alenia Space, with the Spacebus series, and Astrium.

## Molniya Orbit

Geostationary satellites must operate above the equator and therefore appear lower on the horizon as the receiver gets farther from the equator. This will cause problems for extreme northerly latitudes, affecting connectivity and causing multipath interference (caused by signals reflecting off the ground and into the ground antenna).

Thus, for areas close to the North (and South) Pole, a geostationary satellite may appear below the horizon. Therefore, Molniya orbit satellites have been launched, mainly in Russia, to alleviate this problem.

Molniya orbits can be an appealing alternative in such cases. The Molniya orbit is highly inclined, guaranteeing good elevation over selected positions during the northern portion of the orbit. (Elevation is the extent of the satellite's position above the horizon. Thus, a satellite at the horizon has zero elevation and a satellite directly overhead has elevation of 90 degrees.)

The Molniya orbit is designed so that the satellite spends the great majority of its time over the far northern latitudes, during which its ground footprint moves only slightly. Its period is one half day, so that the satellite is available for operation over the targeted region for six to nine hours every second revolution. In this way a constellation of three Molniya satellites (plus in-orbit spares) can provide uninterrupted coverage.

The first satellite of the Molniya series was launched on April 23, 1965 and was used for experimental transmission of TV signals from a Moscow uplink station to downlink stations located in Siberia and the Russian Far East, in Norilsk, Khabarovsk, Magadan and Vladivostok. In November 1967 Soviet engineers created a unique system of national TV network of satellite television, called Orbita, that was based on Molniya satellites.

## Polar Orbit

In the United States, the National Polar-orbiting Operational Environmental Satellite System (NPOESS) was established in 1994 to consolidate the polar satellite operations of NASA (National Aeronautics and Space Administration) NOAA (National Oceanic and Atmospheric Administration). NPOESS manages a number of satellites for various purposes; for example, METSAT for meteorological satellite, EUMETSAT for the European branch of the program, and METOP for meteorological operations.

These orbits are sun synchronous, meaning that they cross the equator at the same local time each day. For example, the satellites in the NPOESS (civilian) orbit will cross the equator, going from south to north, at times 1:30 P.M., 5:30 P.M., and 9:30 P.M.

## Structure

Communications Satellites are usually composed of the following subsystems:

- Communication Payload, normally composed of transponders, antennas, and switching systems.

- Engines used to bring the satellite to its desired orbit.

- A station keeping tracking and stabilization subsystem used to keep the satellite in the right orbit, with its antennas pointed in the right direction, and its power system pointed towards the sun.

- Power subsystem, used to power the Satellite systems, normally composed of solar cells, and batteries that maintain power during solar eclipse.

- Command and Control subsystem, which maintains communications with ground control stations. The ground control Earth stations monitor the satellite performance and control its functionality during various phases of its life-cycle.

The bandwidth available from a satellite depends upon the number of transponders provided by the satellite. Each service (TV, Voice, Internet, radio) requires a different amount of bandwidth for transmission. This is typically known as link budgeting and a network simulator can be used to arrive at the exact value.

## Frequency Allocation for Satellite Systems

Allocating frequencies to satellite services is a complicated process which requires international coordination and planning. This is carried out under the auspices of the International Telecommunication Union (ITU). To facilitate frequency planning, the world is divided into three regions: Region 1: Europe, Africa, what was formerly the Soviet Union, and Mongolia Region 2: North and South America and Greenland Region 3: Asia (excluding region 1 areas), Australia, and the southwest Pacific.

Within these regions, frequency bands are allocated to various satellite services, although a given service may be allocated different frequency bands in different regions. Some of the services provided by satellites are:

- Fixed satellite service (FSS).

- Broadcasting satellite service (BSS).

- Mobile-satellite service.

- Radionavigation-satellite service.

- Meteorological-satellite service.

## Cellular Communication using Satellites

A Satellite communication is a technology that is used to transfer the signals from the transmitter to a receiver with the help of satellites. It can be used in different mobile applications that involve communication with the ships, vehicles and radio broadcasting services. The power and bandwidth of these satellites depend on the specifications like complexity, size and cost.

Satellites can be used in these applications: weather forecasting, satellite-based military applications, global telephones connecting remote areas, global mobile communications, radio and TV broadcasting, and so on.

Satellite services are classified into three types based on the frequency allocation. Frequency allocation is a typical process done with the help of international coordination and planning. Satellite services include fixed satellite service, broadcasting satellite service and mobile satellite service.

## Mobile Satellite Service

Mobile satellite service (MSS) is the term used to describe telecommunication services delivered to or from the mobile users by using the satellites. MSS can be used in remote areas lacking wired networks. Limitations of MSS are availability of line of sight requirement and emerging technologies.

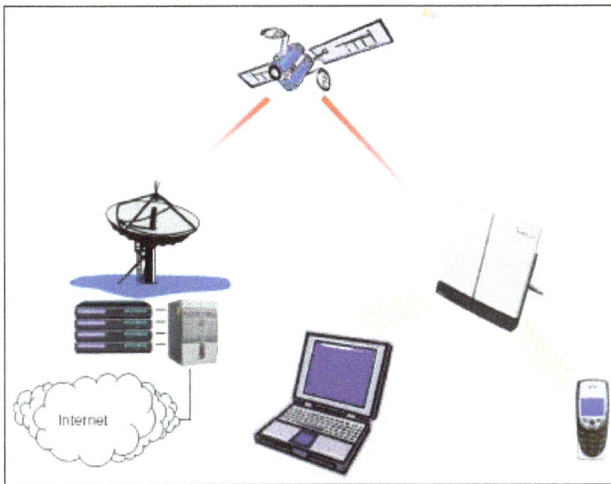

Mobile satellite service (MSS) system.

The basic MSS System comprises of these three segments:

- Space segment.
- User segment.
- Control segment.

Space segment: Space segment is equipped with satellite pay-load equipment. The Pay load is used to enable the ability of the satellite for users in space communication.

User segment: The user segment consists of equipment that transmits and receives the signals from the satellite.

Control segment: The control segment controls the satellite and operations of all internet connections to maintain the bandwidth and adjust power supply and antennas.

The mobile satellite services are classified into the following five types:

- Maritime mobile satellite service (MMSS).

- Land mobile satellite service (LMSS).

- Aeronautical mobile satellite service (AMSS).

- Personal mobiles satellite service (PMSS).

- Broadcast mobile satellite service (BCMSS).

## Maritime Mobile Satellite Service

This service consists of different types of earth stations such as mobile earth station (MES); ship earth station (SES); and communication earth station (CES).This service is mainly used in shipyards and military ships.

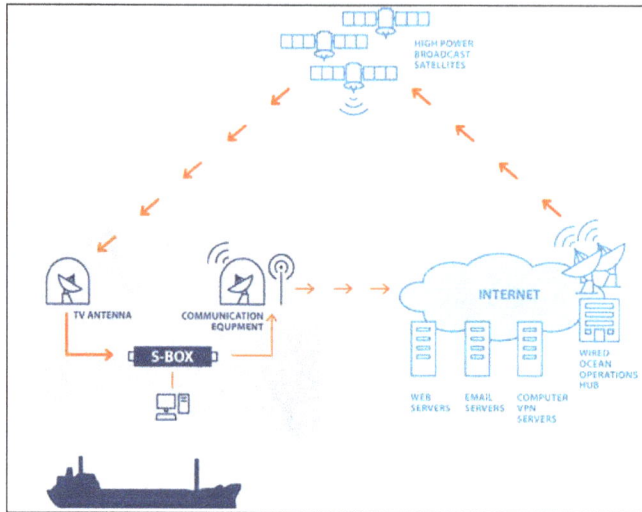

Maritime mobile satellite service.

In this type of service, the mobile earth station located on ships provide commercial and safety communication. MMSS service enables mobile satellite link between the communication earth station and the ship earth station or between two associated ships and other satellite communication stations in all positions in sea or in ports.

A maritime terminal is a portable or fixed on the board ship, whereas the communication earth station is a maritime earth station located at a specified fixed point on the coast to provide a feeder link for MMSS. The ship earth station is a maritime earth station fixed on board ships or other floating objects that provide the communication links with the subscribers onshore via a communication earth station and a communication space craft.

## Land Mobile Satellite Service

The Land mobile satellite service has a mobile earth station located on different types of trains and other transportation systems. This service consists of a personal location beacon terminal that acts as an earth station. This service can be used in different applications such as military applications remote and rural environments.

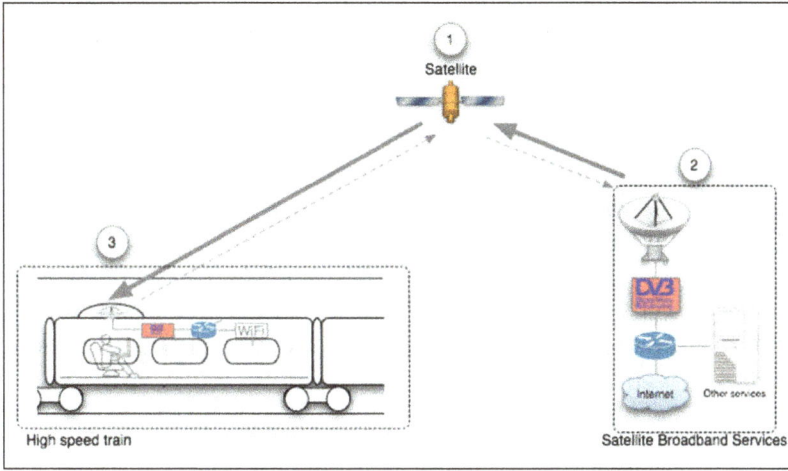

Land mobile satellite service.

LMSS enables a mobile link between the communication earth system and the vehicle's earth system or two or more vehicles' earth stations or two MSC stations. The communication earth station is used as an earth station to be located in a specified fixed point on the coast to provide a feeder link for LMSS.

The VES is a land mobile earth station fixed on the board or rail line to provide a communication link between the terrestrial subscribers through VES and communication spacecraft. The land vehicle or person alerts the service for distrain or safety in the LMSS system.

## Aeronautical Mobile Satellite Service

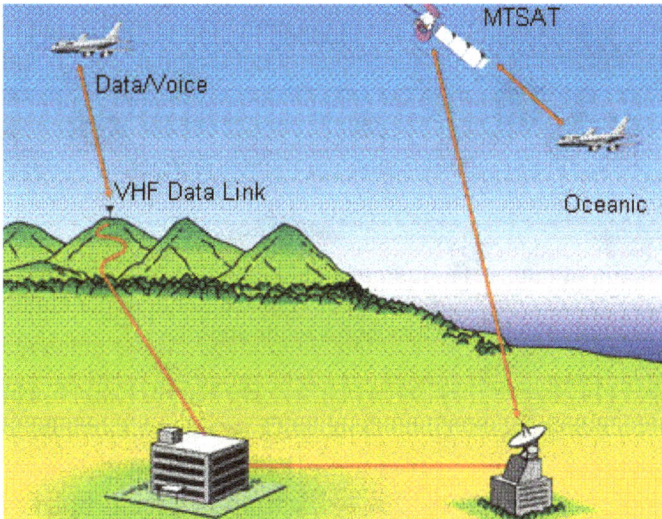

Aeronautical mobile satellite service.

A mobile satellite service in which earth stations are located onboard aircraft, survival aircraft, airplanes and helicopters is known as aeronautical mobile satellite service (AES).This service is also used in business and private communication and traffic control. This service consists of various earth stations like a mobile earth station, an aircraft earth station and a ground earth station.

A special emergency locator terminal which is either fixed or portable onboard is used as earth station and enables the link between the ground earth station and the aircraft earth station. The AES is an aeronautical earth station that is fixed on board to provide a communication link with the subscribers on land via GES and space craft.

This is mostly used in the aircraft applications as it provides safety through the radio communication to control flight locations and the movements of light and the positions of aircraft on land as well.

## Personal Mobile Satellite Service

This is a communication service provided by the satellite for supporting mobile, fixed and broadband communication systems. The satellites can be geo-stationary or non-geo-stationary satellites. This service consists of two earth stations: base earth station and personal earth station. It also consists of a PLB terminal which is used in this service for the coordination of the mobile system.

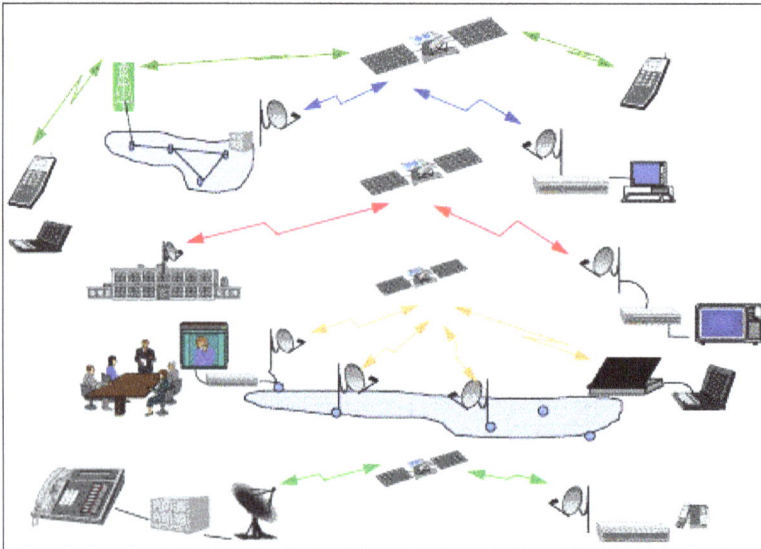

Personal mobile satellite service.

This type of service enables a link between a base earth station and personal earth station, or between a personal earth station, or between an earth base station and two satellites using the same satellite providers. It is a handheld terminal carried by an individual or fixed on board. It provides two communication links for subscribers by satellites through gateways or personal earth station.

## Broadcast Mobile Satellite Service

A broadcast satellite system service is a one-way radio communication solution that transmits signals by earth stations, and retransmits the signals by space stations. The present broadcast mobile satellite service operates at a frequency of 12 GHz.

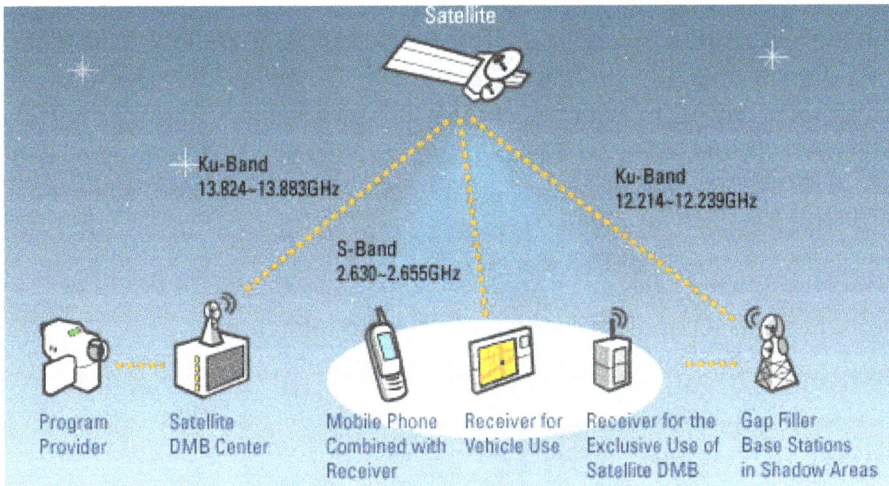

Broadcast mobile satellite service.

The broadcast satellite service system transmits data in three types of broadcasting forms:

- Audio broadcasting.

- Video broadcasting.

- Data broadcasting.

This service is equipped with very small terminals used for transmitting signals from small antennas. This service can be used in applications like ships, airlines and TV broadcasting systems.

## First Generation (1G)

1G refers to the first generation of wireless cellular technology (mobile telecommunications). These are the analog telecommunications standards that were introduced in the 1980s and continued until being replaced by 2G digital telecommunications. The main difference between the two mobile cellular systems (1G and 2G), is that the radio signals used by 1G networks are analog, while 2G networks are digital.

Although both systems use digital signaling to connect the radio towers (which listen to the handsets) to the rest of the telephone system, the voice itself during a call is encoded to digital signals in 2G whereas 1G is only modulated to higher frequency, typically 150 MHz and up. The inherent advantages of digital technology over that of analog meant that 2G networks eventually replaced them almost everywhere.

One such standard is Nordic Mobile Telephone (NMT), used in Nordic countries, Switzerland, the Netherlands, Eastern Europe and Russia. Others include Advanced Mobile Phone System (AMPS) used in North America and Australia, TACS (Total Access Communications System) in the United Kingdom, C-450 in West Germany, Portugal and South Africa, Radiocom 2000 in France, TMA in Spain, and RTMI in Italy. In Japan there were multiple systems. Three standards, TZ-801, TZ-802, and TZ-803 were developed by NTT (Nippon Telegraph and Telephone Corporation), while a competing system operated by Daini Denden Planning, Inc. (DDI) used the Japan Total Access Communications System (JTACS) standard.

The antecedent to 1G technology is the mobile radio telephone.

## Second Generation (2G)

2G (or 2-G) is short for second-generation cellular network. 2G cellular networks were commercially launched on the GSM standard in Finland by Radiolinja (now part of Elisa Oyj) in 1991.

Three primary benefits of 2G networks over their predecessors were that:

- Phone conversations were digitally encrypted.

- Significantly more efficient use of the radio frequency spectrum enabling more users per frequency band.

- Data services for mobile, starting with SMS text messages.

2G technologies enabled the various networks to provide the services such as text messages, picture messages, and MMS (multimedia messages). All text messages sent over 2G are digitally encrypted, allowing the transfer of data in such a way that only the intended receiver can receive and read it.

After 2G was launched, the previous mobile wireless network systems were retroactively dubbed 1G. While radio signals on 1G networks are analog, radio signals on 2G networks are digital. Both systems use digital signaling to connect the radio towers (which listen to the devices) to the rest of the mobile system.

With General Packet Radio Service (GPRS), 2G offers a theoretical maximum transfer speed of 40 kbit/s. With EDGE (Enhanced Data Rates for GSM Evolution), there is a theoretical maximum transfer speed of 384 kbit/s.

The most common 2G technology was the time division multiple access (TDMA)-based GSM, originally from Europe but used in most of the world outside North America. Over 60 GSM operators were also using CDMA2000 in the 450 MHz frequency band (CDMA450) by 2010.

## GSM

The Global System for Mobile Communications (GSM) is a standard developed by the European Telecommunications Standards Institute (ETSI) to describe the protocols for second-generation (2G) digital cellular networks used by mobile devices such as mobile phones and tablets. It was first deployed in Finland in December 1991. By the mid-2010s, it became a global standard for mobile communications achieving over 90% market share, and operating in over 193 countries and territories.

2G networks developed as a replacement for first generation (1G) analog cellular networks. The GSM standard originally described a digital, circuit-switched network optimized for full duplex voice telephony. This expanded over time to include data communications, first by circuit-switched transport, then by packet data transport via General Packet Radio Service (GPRS), and Enhanced Data Rates for GSM Evolution (EDGE).

Subsequently, the 3GPP developed third-generation (3G) UMTS standards, followed by fourth-generation (4G) LTE Advanced standards, which do not form part of the ETSI GSM standard.

"GSM" is a trade mark owned by the GSM Association. It may also refer to the (initially) most common voice codec used, Full Rate.

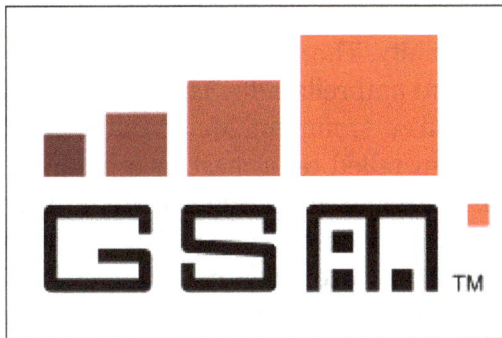

The GSM logo is used to identify compatible devices and equipment. The dots symbolize three clients in the home network and one roaming client.

## Technical Details

## Network Structure

The network is structured into several discrete sections:

- Base station subsystem: The base stations and their controllers.

- Network and Switching Subsystem: The part of the network most similar to a fixed network, sometimes just called the "core network".

- GPRS Core Network: The optional part which allows packet-based Internet connections.

- Operations support system (OSS): Network maintenance.

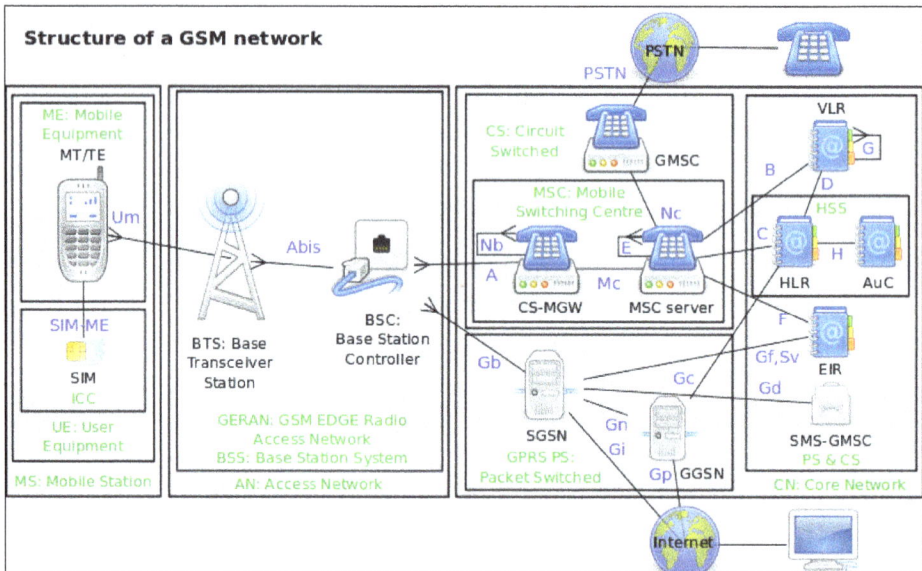

## Base Station Subsystem

GSM is a cellular network, which means that cell phones connect to it by searching for cells in the immediate vicinity. There are five different cell sizes in a GSM network macro, micro, pico, femto, and umbrella cells. The coverage area of each cell varies according to the implementation environment. Macro cells can be regarded as cells where the base station antenna is installed on a mast or a building above average rooftop level. Micro cells are cells whose antenna height is under average rooftop level; they are typically used in urban areas. Picocells are small cells whose coverage diameter is a few dozen meters; they are mainly used indoors. Femtocells are cells designed for use in residential or small business environments and connect to the service provider's network via a broadband internet connection. Umbrella cells are used to cover shadowed regions of smaller cells and fill in gaps in coverage between those cells.

Cell horizontal radius varies depending on antenna height, antenna gain, and propagation conditions from a couple of hundred meters to several tens of kilometers. The longest distance the GSM specification supports in practical use is 35 kilometres (22 mi). There are also several implementations of the concept of an extended cell,where the cell radius could be double or even more, depending on the antenna system, the type of terrain, and the timing advance.

Indoor coverage is also supported by GSM and may be achieved by using an indoor pi-cocell base station, or an indoor repeater with distributed indoor antennas fed through power splitters, to deliver the radio signals from an antenna outdoors to the separate indoor distributed antenna system. These are typically deployed when significant call capacity is needed indoors, like in shopping centers or airports. However, this is not a prerequisite, since indoor coverage is also provided by in-building penetration of the radio signals from any nearby cell.

## GSM Carrier Frequencies

GSM networks operate in a number of different carrier frequency ranges (separated into GSM frequency ranges for 2G and UMTS frequency bands for 3G), with most 2G GSM networks operating in the 900 MHz or 1800 MHz bands. Where these bands were already allocated, the 850 MHz and 1900 MHz bands were used instead (for example in Canada and the United States). In rare cases the 400 and 450 MHz frequency bands are assigned in some countries because they were previously used for first-generation systems.

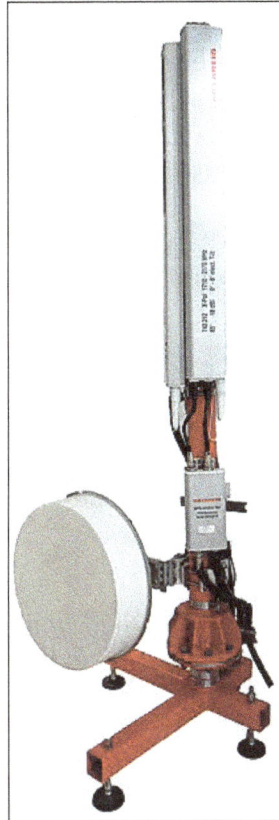

GSM cell site antennas.

For comparison most 3G networks in Europe operate in the 2100 MHz frequency band. For more information on worldwide GSM frequency usage.

Regardless of the frequency selected by an operator, it is divided into timeslots for individual phones. This allows eight full-rate or sixteen half-rate speech channels per radio frequency. These eight radio timeslots (or burst periods) are grouped into a TDMA frame. Half-rate channels use alternate frames in the same timeslot. The channel data rate for all 8 channels is 270.833 Kbit/s, and the frame duration is 4.615 ms.

The transmission power in the handset is limited to a maximum of 2 watts in GSM 850/900 and 1 watt in GSM 1800/1900.

## Voice Codecs

GSM has used a variety of voice codecs to squeeze 3.1 kHz audio into between 7 and 13 kbit/s. Originally, two codecs, named after the types of data channel they were allocated, were used, called Half Rate (6.5 kbit/s) and Full Rate (13 kbit/s). These used a system based on linear predictive coding (LPC). In addition to being efficient with bitrates, these codecs also made it easier to identify more important parts of the audio, allowing the air interface layer to prioritize and better protect these parts of the signal. GSM was further enhanced in 1997 with the enhanced full rate (EFR) codec, a 12.2 kbit/s codec that uses a full-rate channel. Finally, with the development of UMTS, EFR was refactored into a variable-rate codec called AMR-Narrowband, which is high quality and robust against interference when used on full-rate channels, or less robust but still relatively high quality when used in good radio conditions on half-rate channel.

## Subscriber Identity Module (SIM)

One of the key features of GSM is the Subscriber Identity Module, commonly known as a SIM card. The SIM is a detachable smart card containing the user's subscription information and phone book. This allows the user to retain his or her information after switching handsets. Alternatively, the user can change operators while retaining the handset simply by changing the SIM. Some operators will block this by allowing the phone to use only a single SIM, or only a SIM issued by them; this practice is known as SIM locking.

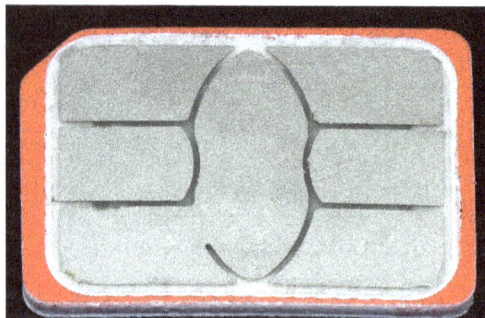

A nano sim used in mobile phones.

## Phone Locking

Sometimes mobile network operators restrict handsets that they sell for exclusive use in their own network. This is called SIM locking and is implemented by a software feature of the phone. A subscriber may usually contact the provider to remove the lock for a fee, utilize private services to remove the lock, or use software and websites to unlock the handset themselves. It is possible to hack past a phone locked by a network operator.

In some countries (e.g., Bangladesh, Belgium, Brazil, Canada, Chile, Germany, Hong Kong, India, Iran, Lebanon, Malaysia, Nepal, Norway, Pakistan, Poland, Singapore, South Africa, Thailand) all phones are sold unlocked due to the abundance of dual SIM handsets and operators.

## GSM Security

GSM was intended to be a secure wireless system. It has considered the user authentication using a pre-shared key and challenge-response, and over-the-air encryption. However, GSM is vulnerable to different types of attack, each of them aimed at a different part of the network.

The development of UMTS introduced an optional Universal Subscriber Identity Module (USIM), that uses a longer authentication key to give greater security, as well as mutually authenticating the network and the user, whereas GSM only authenticates the user to the network (and not vice versa). The security model therefore offers confidentiality and authentication, but limited authorization capabilities, and no non-repudiation.

GSM uses several cryptographic algorithms for security. The A5/1, A5/2, and A5/3 stream ciphers are used for ensuring over-the-air voice privacy. A5/1 was developed first and is a stronger algorithm used within Europe and the United States; A5/2 is weaker and used in other countries. Serious weaknesses have been found in both algorithms: it is possible to break A5/2 in real-time with a ciphertext-only attack, and in January 2007, The Hacker's Choice started the A5/1 cracking project with plans to use FPGAs that allow A5/1 to be broken with a rainbow table attack. The system supports multiple algorithms so operators may replace that cipher with a stronger one.

Since 2000 different efforts have been made in order to crack the A5 encryption algorithms. Both A5/1 and A5/2 algorithms have been broken, and their cryptanalysis has been revealed in the literature. As an example, Karsten Nohl developed a number of rainbow tables (static values which reduce the time needed to carry out an attack) and have found new sources for known plaintext attacks. He said that it is possible to build "a full GSM interceptor from open-source components" but that they had not done so because of legal concerns. Nohl claimed that he was able to intercept voice and text conversations by impersonating another user to listen to voicemail, make calls, or send text messages using a seven-year-old Motorola cellphone and decryption software available for free online.

GSM uses General Packet Radio Service (GPRS) for data transmissions like browsing the web. The most commonly deployed GPRS ciphers were publicly broken in 2011.

The researchers revealed flaws in the commonly used GEA/1 and GEA/2 ciphers and published the open-source "gprsdecode" software for sniffing GPRS networks. They also noted that some carriers do not encrypt the data (i.e., using GEA/0) in order to detect the use of traffic or protocols they do not like (e.g., Skype), leaving customers unprotected. GEA/3 seems to remain relatively hard to break and is said to be in use on some more modern networks. If used with USIM to prevent connections to fake base stations and downgrade attacks, users will be protected in the medium term, though migration to 128-bit GEA/4 is still recommended.

## Standards Information

The GSM systems and services are described in a set of standards governed by ETSI, where a full list is maintained.

## GSM Open-source Software

Several open-source software projects exist that provide certain GSM features:

- Gsmd daemon by Openmoko.

- OpenBTS develops a Base transceiver station.

- The GSM Software Project aims to build a GSM analyzer for less than $1,000.

- OsmocomBB developers intend to replace the proprietary baseband GSM stack with a free software implementation.

- YateBTS develops a Base transceiver station.

## Issues with Patents and Open-source

Patents remain a problem for any open-source GSM implementation, because it is not possible for GNU or any other free software distributor to guarantee immunity from all lawsuits by the patent holders against the users. Furthermore, new features are being added to the standard all the time which means they have patent protection for a number of years.

The original GSM implementations from 1991 may now be entirely free of patent encumbrances, however patent freedom is not certain due to the United States' "first to invent" system that was in place until 2012. The "first to invent" system, coupled with "patent term adjustment" can extend the life of a U.S. patent far beyond 20 years from its priority date. It is unclear at this time whether OpenBTS will be able to implement features of that initial specification without limit. As patents subsequently expire, however, those features can be added into the open-source version. As of 2011, there have been no lawsuits against users of OpenBTS over GSM use.

## CDMA

CDMA (Code-Division Multiple Access) refers to any of several protocols used in second-generation (2G) and third-generation (3G) wireless communications. As the term implies, CDMA is a form of multiplexing, which allows numerous signals to occupy a single transmission channel, optimizing the use of available bandwidth. The technology is used in ultra-high-frequency (UHF) cellular telephone systems in the 800-MHz and 1.9-GHz bands.

CDMA employs analog-to-digital conversion (ADC) in combination with spread spectrum technology. Audio input is first digitized into binary elements. The frequency of the transmitted signal is then made to vary according to a defined pattern (code), so it can be intercepted only by a receiver whose frequency response is programmed with the same code, so it follows exactly along with the transmitter frequency. There are trillions of possible frequency-sequencing codes, which enhance privacy and makes cloning difficult.

The CDMA channel is nominally 1.23 MHz wide. CDMA networks use a scheme called soft handoff, which minimizes signal breakup as a handset passes from one cell to another. The combination of digital and spread-spectrum modes supports several times as many signals per unit bandwidth as analog modes. CDMA is compatible with other cellular technologies; this allows for nationwide roaming. The original CDMA standard, also known as CDMA One, offers a transmission speed of only up to 14.4 Kbps in its single channel form and up to 115 Kbps in an eight-channel form. CDMA2000 and Wideband CDMA deliver data many times faster.

The CDMA2000 family of standards includes 1xRTT, EV-DO Rev 0, EV-DO Rev A and EV-DO Rev B (renamed Ultra Mobile Broadband -- UMB). People often confuse CDMA2000 (a family of standards supported by Verizon and Sprint) with CDMA (the physical layer multiplexing scheme).

### cdmaOne

Interim Standard 95 (IS-95) was the first ever CDMA-based digital cellular technology. It was developed by Qualcomm and later adopted as a standard by the Telecommunications Industry Association in TIA/EIA/IS-95 release published in 1995. The proprietary name for IS-95 is cdmaOne.

It is a 2G mobile telecommunications standard that uses CDMA, a multiple access scheme for digital radio, to send voice, data and signaling data (such as a dialed telephone number) between mobile telephones and cell sites.

CDMA or "code division multiple access" is a digital radio system that transmits streams of bits (PN codes). CDMA permits several radios to share the same frequencies. Unlike TDMA "time division multiple access", a competing system used in 2G GSM, all radios can be active all the time, because network capacity does not directly limit the number of active radios. Since larger numbers of phones can be served by smaller numbers of cell-sites, CDMA-based standards have a significant economic advantage over TDMA-based standards, or the oldest cellular standards that used frequency-division multiplexing.

In North America, the technology competed with Digital AMPS (IS-136, a TDMA technology). It is now being supplanted by IS-2000 (CDMA2000), a later CDMA-based standard.

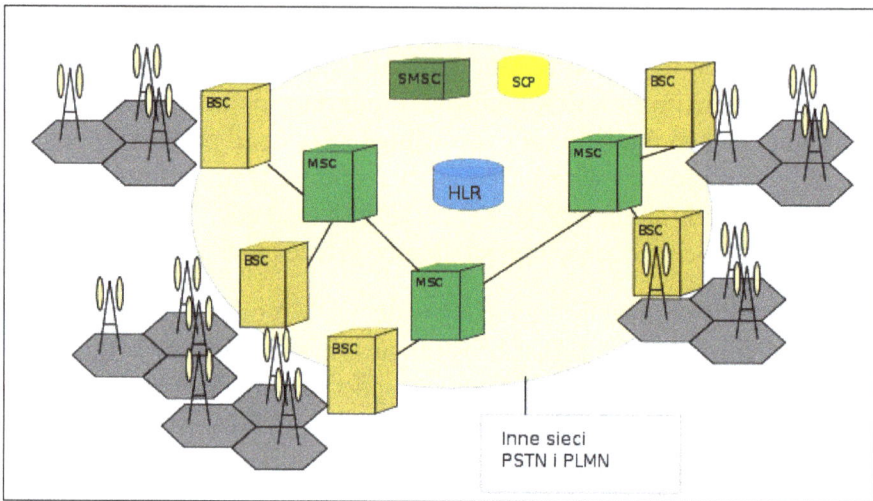

cdmaOne network structure.

## Protocol Revisions

cdmaOne's technical history is reflective of both its birth as a Qualcomm internal project, and the world of then-unproven competing digital cellular standards under which it was developed. The term IS-95 generically applies to the earlier set of protocol revisions, namely P_REV's one through five.

P_REV=1 was developed under an ANSI standards process with documentation reference J-STD-008. J-STD-008, published in 1995, was only defined for the then-new North American PCS band (Band Class 1, 1900 MHz). The term IS-95 properly refers to P_REV=1, developed under the Telecommunications Industry Association (TIA) standards process, for the North American cellular band (Band Class 0, 800 MHz) under roughly the same time frame. IS-95 offered interoperation (including handoff) with the analog cellular network. For digital operation, IS-95 and J-STD-008 have most technical details in common. The immature style and structure of both documents are highly reflective of the "standardizing" of Qualcomm's internal project.

P_REV=2 is termed Interim Standard 95A (IS-95A). IS-95A was developed for Band Class 0 only, as in incremental improvement over IS-95 in the TIA standards process.

P_REV=3 is termed Technical Services Bulletin 74 (TSB-74). TSB-74 was the next incremental improvement over IS-95A in the TIA standards process.

P_REV=4 is termed Interim Standard 95B (IS-95B) Phase I, and P_REV=5 is termed Interim Standard 95B (IS-95B) Phase II. The IS-95B standards track provided for a merging of the TIA and ANSI standards tracks under the TIA, and was the first document that provided for interoperation of IS-95 mobile handsets in both band classes (dual-band operation). P_REV=4 was by far the most popular variant of IS-95, with P_REV=5 only seeing minimal uptake in South Korea.

P_REV=6 and beyond fall under the CDMA2000 umbrella. Besides technical improvements, the IS-2000 documents are much more mature in terms of layout and content. They also provide backwards-compatibility to IS-95.

## Protocol Details

The IS-95 standards describe an air interface, a set of protocols used between mobile units and the network. IS-95 is widely described as a three-layer stack, where L1 corresponds to the physical (PHY) layer, L2 refers to the Media Access Control (MAC) and Link-Access Control (LAC) sublayers, and L3 to the call-processing state machine.

### Physical Layer

IS-95 defines the transmission of signals in both the forward (network-to-mobile) and reverse (mobile-to-network) directions.

In the forward direction, radio signals are transmitted by base stations (BTS's). Every BTS is synchronized with a GPS receiver so transmissions are tightly controlled in time. All forward transmissions are QPSK with a chip rate of 1,228,800 per second. Each signal is spread with a Walsh code of length 64 and a pseudo-random noise code (PN code) of length 215, yielding a PN roll-over period of $\frac{80}{3}$ ms.

For the reverse direction, radio signals are transmitted by the mobile. Reverse link transmissions are OQPSK in order to operate in the optimal range of the mobile's power amplifier. Like the forward link, the chip rate is 1,228,800 per second and signals are spread with Walsh codes and the pseudo-random noise code, which is also known as a Short Code.

### Forward Broadcast Channels

Every BTS dedicates a significant amount of output power to a pilot channel, which is an unmodulated PN sequence (in other words, spread with Walsh code 0). Each BTS sector in the network is assigned a PN offset in steps of 64 chips. There is no data carried on the forward pilot. With its strong autocorrelation function, the forward pilot allows mobiles to determine system timing and distinguish different BTS's for handoff.

When a mobile is "searching", it is attempting to find pilot signals on the network by tuning to particular radio frequencies, and performing a cross-correlation across all possible PN phases. A strong correlation peak result indicates the proximity of a BTS.

Other forward channels, selected by their Walsh code, carry data from the network to the mobiles. Data consists of network signaling and user traffic. Generally, data to be transmitted is divided into frames of bits. A frame of bits is passed through a convolutional encoder, adding forward error correction redundancy, generating a frame of symbols. These symbols are then spread with the Walsh and PN sequences and transmitted.

BTSs transmit a sync channel spread with Walsh code 32. The sync channel frame is $\frac{80}{3}$ ms long, and its frame boundary is aligned to the pilot. The sync channel continually transmits a single message, the Sync Channel Message, which has a length and content dependent on the P_REV. The message is transmitted 32 bits per frame, encoded to 128 symbols, yielding a rate of 1200 bit/s. The Sync Channel Message contains information about the network, including the PN offset used by the BTS sector.

Once a mobile has found a strong pilot channel, it listens to the sync channel and decodes a Sync Channel Message to develop a highly accurate synchronization to system time. At this point the mobile knows whether it is roaming, and that it is "in service".

BTSs transmit at least one, and as many as seven, paging channels starting with Walsh code 1. The paging channel frame time is 20 ms, and is time aligned to the IS-95 system (i.e. GPS) 2-second roll-over. There are two possible rates used on the paging channel: 4800 bit/s or 9600 bit/s. Both rates are encoded to 19200 symbols per second.

The paging channel contains signaling messages transmitted from the network to all idle mobiles. A set of messages communicate detailed network overhead to the mobiles, circulating this information while the paging channel is free. The paging channel also carries higher-priority messages dedicated to setting up calls to and from the mobiles.

When a mobile is idle, it is mostly listening to a paging channel. Once a mobile has parsed all the network overhead information, it registers with the network, and then optionally enters slotted-mode.

**Forward Traffic Channels**

The Walsh space not dedicated to broadcast channels on the BTS sector is available for traffic channels. These channels carry the individual voice and data calls supported by IS-95. Like the paging channel, traffic channels have a frame time of 20ms.

Since voice and user data are intermittent, the traffic channels support variable-rate operation. Every 20 ms frame may be transmitted at a different rate, as determined by the service in use (voice or data). P_REV=1 and P_REV=2 supported rate set 1, providing a rate of 1200, 2400, 4800, or 9600 bit/s. P_REV=3 and beyond also provided rate set 2, yielding rates of 1800, 3600, 7200, or 14400 bit/s.

For voice calls, the traffic channel carries frames of vocoder data. A number of different vocoders are defined under IS-95, the earlier of which were limited to rate set 1, and were responsible for some user complaints of poor voice quality. More sophisticated vocoders, taking advantage of modern DSPs and rate set 2, remedied the voice quality situation and are still in wide use in 2005.

The mobile receiving a variable-rate traffic frame does not know the rate at which the frame was transmitted. Typically, the frame is decoded at each possible rate, and using the quality metrics of the Viterbi decoder, the correct result is chosen.

Traffic channels may also carry circuit-switch data calls in IS-95. The variable-rate traffic frames are generated using the IS-95 Radio Link Protocol (RLP). RLP provides a mechanism to improve the performance of the wireless link for data. Where voice calls might tolerate the dropping of occasional 20 ms frames, a data call would have unacceptable performance without RLP.

Under IS-95B P_REV=5, it was possible for a user to use up to seven supplemental "code" (traffic) channels simultaneously to increase the throughput of a data call. Very few mobiles or networks ever provided this feature, which could in theory offer 115200 bit/s to a user.

## Block Interleaver

After convolution coding and repetition, symbols are sent to a 20 ms block interleaver, which is a 24 by 16 array.

## Capacity

IS-95 and its use of CDMA techniques, like any other communications system, have their throughput limited according to Shannon's theorem. Accordingly, capacity improves with SNR and bandwidth. IS-95 has a fixed bandwidth, but fares well in the digital world because it takes active steps to improve SNR.

With CDMA, signals that are not correlated with the channel of interest (such as other PN offsets from adjacent cellular base stations) appear as noise, and signals carried on other Walsh codes (that are properly time aligned) are essentially removed in the de-spreading process. The variable-rate nature of traffic channels provide lower-rate frames to be transmitted at lower power causing less noise for other signals still to be correctly received. These factors provide an inherently lower noise level than other cellular technologies allowing the IS-95 network to squeeze more users into the same radio spectrum.

Active (slow) power control is also used on the forward traffic channels, where during a call, the mobile sends signaling messages to the network indicating the quality of the signal. The network will control the transmitted power of the traffic channel to keep the signal quality just good enough, thereby keeping the noise level seen by all other users to a minimum.

The receiver also uses the techniques of the rake receiver to improve SNR as well as perform soft handoff.

### Layer 2

Once a call is established, a mobile is restricted to using the traffic channel. A frame format is defined in the MAC for the traffic channel that allows the regular voice (vocoder) or data (RLP) bits to be multiplexed with signaling message fragments. The signaling message fragments are pieced together in the LAC, where complete signaling messages are passed on to Layer 3.

## Third Generation (3G)

3G is the third generation of wireless mobile telecommunications technology. It is the upgrade for 2G and 2.5G GPRS networks, for faster data transfer speed. This is based on a set of standards used for mobile devices and mobile telecommunications use services and networks that comply with the International Mobile Telecommunications-2000 (IMT-2000) specifications by the International Telecommunication Union. 3G finds application in wireless voice telephony, mobile Internet access, fixed wireless Internet access, video calls and mobile TV.

3G telecommunication networks support services that provide an information transfer rate of at least 144 Kbit/s. Later 3G releases, often denoted 3.5G and 3.75G, also provide mobile broadband access of several Mbit/s to smartphones and mobile modems in laptop computers. This ensures it can be applied to wireless voice telephony, mobile Internet access, fixed wireless Internet access, video calls and mobile TV technologies.

A new generation of cellular standards has appeared approximately every tenth year since 1G systems were introduced in 1979 and the early to mid-1980s. Each generation is characterized by new frequency bands, higher data rates and non–backward-compatible transmission technology. The first commercial 3G networks were introduced in 2000.

Several telecommunications companies market wireless mobile Internet services as 3G, indicating that the advertised service is provided over a 3G wireless network. Services advertised as 3G are required to meet IMT-2000 technical standards, including standards for reliability and speed (data transfer rates). To meet the IMT-2000 standards, a system is required to provide peak data rates of at least 144 Kbit/s. However, many services advertised as 3G provide higher speed than the minimum technical requirements for a 3G service. Recent 3G releases, often denoted 3.5G and 3.75G, also provide mobile broadband access of several Mbit/s to smartphones and mobile modems in laptop computers.

The following standards are typically branded 3G:

- The UMTS (Universal Mobile Telecommunications System) system, first offered in 2001, standardized by 3GPP, used primarily in Europe, Japan, China (however with a different radio interface) and other regions predominated by GSM (Global Systems for Mobile) 2G system infrastructure. The cell phones are typically UMTS and GSM hybrids. Several radio interfaces are offered, sharing the same infrastructure:

  ○ The original and most widespread radio interface is called W-CDMA (Wideband Code Division Multiple Access).

  ○ The TD-SCDMA radio interface was commercialized in 2009 and is only offered in China.

  ○ The latest UMTS release, HSPA+, can provide peak data rates up to 56 Mbit/s in the downlink in theory (28 Mbit/s in existing services) and 22 Mbit/s in the uplink.

- The CDMA2000 system, first offered in 2002, standardized by 3GPP2, used especially in North America and South Korea, sharing infrastructure with the IS-95 2G standard. The cell phones are typically CDMA2000 and IS-95 hybrids. The latest release EVDO Rev B offers peak rates of 14.7 Mbit/s downstream.

The above systems and radio interfaces are based on spread spectrum radio transmission technology. While the GSM EDGE standard ("2.9G"), DECT cordless phones and Mobile WiMAX standards formally also fulfill the IMT-2000 requirements and are approved as 3G standards by ITU, these are typically not branded 3G, and are based on completely different technologies.

The following common standards comply with the IMT2000/3G standard:

- EDGE, a revision by the 3GPP organization to the older 2G GSM based transmission methods, utilizing the same switching nodes, base station sites and frequencies as GPRS, but new base station and cellphone RF circuits. It is based on the three times as efficient 8PSK modulation scheme as supplement to the original GMSK modulation scheme. EDGE is still used extensively due to its ease of upgrade from existing 2G GSM infrastructure and cell-phones.

- ◦ EDGE combined with the GPRS 2.5G technology is called EGPRS, and allows peak data rates in the order of 200 Kbit/s, just as the original UMTS WCDMA versions, and thus formally fulfills the IMT2000 requirements on 3G systems. However, in practice EDGE is seldom marketed as a 3G system, but a 2.9G system. EDGE shows slightly better system spectral efficiency than the original UMTS and CDMA2000 systems, but it is difficult to reach much higher peak data rates due to the limited GSM spectral bandwidth of 200 kHz, and it is thus a dead end.

- ◦ EDGE was also a mode in the IS-136 TDMA system, today ceased.

- ◦ Evolved EDGE, the latest revision, has peaks of 1 Mbit/s downstream and 400 Kbit/s upstream, but is not commercially used.

- The Universal Mobile Telecommunications System created and revised by the 3GPP. The family is a full revision from GSM in terms of encoding methods and hardware, although some GSM sites can be retrofitted to broadcast in the UMTS/W-CDMA format.

  - ◦ W-CDMA is the most common deployment, commonly operated on the 2,100 MHz band. A few others use the 10, 900 and 1,900 MHz bands.

    - ▫ HSPA is an amalgamation of several upgrades to the original W-CDMA standard and offers speeds of 14.4 Mbit/s down and 5.76 Mbit/s up. HSPA is backward-compatible with and uses the same frequencies as W-CDMA.

    - ▫ HSPA+, a further revision and upgrade of HSPA, can provide theoretical peak data rates up to 168 Mbit/s in the downlink and 22 Mbit/s in the uplink, using a combination of air interface improvements as well as multi-carrier HSPA and MIMO. Technically though, MIMO and DC-HSPA can be used without the "+" enhancements of HSPA+.

- The CDMA2000 system, or IS-2000, including CDMA2000 1x and CDMA2000 High Rate Packet Data (or EVDO), standardized by 3GPP2 (differing from the 3GPP), evolving from the original IS-95 CDMA system, is used especially in North America, China, India, Pakistan, Japan, South Korea, Southeast Asia, Europe and Africa.

  - ◦ CDMA2000 1x Rev. E has an increased voice capacity (in excess of three times) compared to Rev. 0 EVDO Rev. B offers downstream peak rates of 14.7 Mbit/s while Rev. C enhanced existing and new terminal user experience.

While DECT cordless phones and Mobile WiMAX standards formally also fulfill the IMT-2000 requirements, they are not usually considered due to their rarity and unsuitability for usage with mobile phones.

## Break-up of 3G Systems

The 3G (UMTS and CDMA2000) research and development projects started in 1992. In 1999, ITU approved five radio interfaces for IMT-2000 as a part of the ITU-R M.1457 Recommendation; WiMAX was added in 2007.

There are evolutionary standards (EDGE and CDMA) that are backward-compatible extensions to pre-existing 2G networks as well as revolutionary standards that require all-new network hardware and frequency allocations. The cell phones use UMTS in combination with 2G GSM standards and bandwidths, but do not support EDGE. The latter group is the UMTS family, which consists of standards developed for IMT-2000, as well as the independently developed standards DECT and WiMAX, which were included because they fit the IMT-2000 definition.

While EDGE fulfills the 3G specifications, most GSM/UMTS phones report EDGE ("2.75G") and UMTS ("3G") functionality.

## Features

### Data Rates

ITU has not provided a clear definition of the data rate that users can expect from 3G equipment or providers. Thus users sold 3G service may not be able to point to a standard and say that the rates it specifies are not being met. While stating in commentary that "it is expected that IMT-2000 will provide higher transmission rates: a minimum data rate of 2 Mbit/s for stationary or walking users, and 348 Kbit/s in a moving vehicle, "the ITU does not actually clearly specify minimum required rates, nor required average rates, nor what modes of the interfaces qualify as 3G, so various data rates are sold as '3G' in the market.

In market implementation, 3G downlink data speeds defined by telecom service providers vary depending on the underlying technology deployed; up to 384kbit/s for WCDMA, up to 7.2Mbit/sec for HSPA and a theoretical maximum of 21.6 Mbit/s for HSPA+ (technically 3.5G, but usually clubbed under the trade name of 3G). Compare data speeds with 3.5G and 4G.

## Security

3G networks offer greater security than their 2G predecessors. By allowing the UE (User Equipment) to authenticate the network it is attaching to, the user can be sure the network is the intended one and not an impersonator. 3G networks use the KASUMI block cipher instead of the older A5/1 stream cipher. However, a number of serious weaknesses in the KASUMI cipher have been identified.

In addition to the 3G network infrastructure security, end-to-end security is offered when application frameworks such as IMS are accessed, although this is not strictly a 3G property.

## Applications of 3G

The bandwidth and location information available to 3G devices gives rise to applications not previously available to mobile phone users.

## Evolution

Both 3GPP and 3GPP2 are working on the extensions to 3G standards that are based on an all-IP network infrastructure and using advanced wireless technologies such as MIMO. These specifications already display features characteristic for IMT-Advanced (4G), the successor of 3G. However, falling short of the bandwidth requirements for 4G (which is 1 Gbit/s for stationary and 100 Mbit/s for mobile operation), these standards are classified as 3.9G or Pre-4G.

3GPP plans to meet the 4G goals with LTE Advanced, whereas Qualcomm has halted development of UMB in favor of the LTE family.

On 14 December 2009, Telia Sonera announced in an official press release that "We are very proud to be the first operator in the world to offer our customers 4G services. "With the launch of their LTE network, initially they are offering pre-4G (or beyond 3G) services in Stockholm, Sweden and Oslo, Norway.

## CDMA2000

CDMA2000 (also known as C2K or IMT Multi-Carrier (IMT-MC)) is a family of 3G mobile technology standards for sending voice, data, and signaling data between mobile phones and cell sites. It is developed by 3GPP2 as a backwards-compatible successor to second-generation cdmaOne (IS-95) set of standards and used especially in North America and South Korea.

CDMA2000 compares to UMTS, a competing set of 3G standards, which is developed by 3GPP and used in Europe, Japan, and China.

The name CDMA2000 denotes a family of standards that represent the successive, evolutionary stages of the underlying technology. These are:

- Voice: CDMA2000 1xRTT, 1X Advanced.

- Data: CDMA2000 1xEV-DO (Evolution-Data Optimized): Release 0, Revision A, Revision B, Ultra Mobile Broadband (UMB).

All are approved radio interfaces for the ITU's IMT-2000. In the United States, CDMA2000 is a registered trademark of the Telecommunications Industry Association (TIA-USA).

## 1X

CDMA2000 1X (IS-2000), also known as 1x and 1xRTT, is the core CDMA2000 wireless air interface standard. The designation "1x", meaning 1 times radio transmission technology, indicates the same radio frequency (RF) bandwidth as IS-95: a duplex pair of 1.25 MHz radio channels. 1xRTT almost doubles the capacity of IS-95 by adding 64 more traffic channels to the forward link, orthogonal to (in quadrature with) the original set of 64. The 1X standard supports packet data speeds of up to 153 Kbit/s with real world data transmission averaging 80–100 Kbit/s in most commercial applications. IMT-2000 also made changes to the data link layer for greater use of data services, including medium and link access control protocols and QoS. The IS-95 data link layer only provided "best efforts delivery" for data and circuit switched channel for voice (i.e., a voice frame once every 20 ms).

## 1xEV-DO

CDMA2000 1xEV-DO (Evolution-Data Optimized), often abbreviated as EV-DO or EV, is a telecommunications standard for the wireless transmission of data through radio signals, typically for broadband Internet access. It uses multiplexing techniques including code division multiple access (CDMA) as well as time-division access to maximize both individual user's throughput and the overall system throughput. It is standardized (IS-856) by 3rd Generation Partnership Project 2 (3GPP2) as part of the CDMA2000 family of standards and has been adopted by many mobile phone service providers around the world – particularly those previously employing CDMA networks.

## 1X Advanced

1X Advanced (Rev.E) is the evolution of CDMA2000 1X. It provides up to four times the capacity and 70% more coverage compared to 1X.

## Networks

The CDMA Development Group states that, as of April 2014, there are 314 operators in 118 countries offering CDMA2000 1X and/or 1xEV-DO service.

# Fourth Generation (4G)

4G is the fourth generation of broadband cellular network technology, succeeding 3G. A 4G system must provide capabilities defined by ITU in IMT Advanced. Potential and current applications include amended mobile web access, IP telephony, gaming services, high-definition mobile TV, video conferencing, and 3D television.

The first-release Long Term Evolution (LTE) standard was commercially deployed in Oslo, Norway, and Stockholm, Sweden in 2009, and has since been deployed throughout most parts of the world. It has, however, been debated whether first-release versions should be considered 4G LTE.

## Backgrounds of 4G

In the field of mobile communications, a "generation" generally refers to a change in the fundamental nature of the service, non-backwards-compatible transmission technology, higher peak bit rates, new frequency bands, wider channel frequency bandwidth in Hertz, and higher capacity for many simultaneous data transfers (higher system spectral efficiency in bit/second/Hertz/site).

New mobile generations have appeared about every ten years since the first move from 1981 analog (1G) to digital (2G) transmission in 1992. This was followed, in 2001, by 3G multi-media support, spread spectrum transmission and, at least, 200 Kbit/s peak bit rate, in 2011/2012 to be followed by "real" 4G, which refers to all-Internet Protocol (IP) packet-switched networks giving mobile ultra-broadband (gigabit speed) access.

While the ITU has adopted recommendations for technologies that would be used for future global communications, they do not actually perform the standardization or development work themselves, instead relying on the work of other standard bodies such as IEEE, WiMAX Forum, and 3GPP.

In the mid-1990s, the ITU-R standardization organization released the IMT-2000 requirements as a framework for what standards should be considered 3G systems, requiring 200 Kbit/s peak bit rate. In 2008, ITU-R specified the IMT Advanced (International Mobile Telecommunications Advanced) requirements for 4G systems.

The fastest 3G-based standard in the UMTS family is the HSPA+ standard, which is commercially available since 2009 and offers 28 Mbit/s downstream (22 Mbit/s upstream) without MIMO, i.e. only with one antenna, and in 2011 accelerated up to 42 Mbit/s peak bit rate downstream using either DC-HSPA+ (simultaneous use of two 5 MHz UMTS carriers) or 2x2 MIMO. In theory speeds up to 672 Mbit/s are possible, but have not been deployed yet. The fastest 3G-based standard in the CDMA2000 family is the EV-DO Rev. B, which is available since 2010 and offers 15.67 Mbit/s downstream.

## Frequencies for 4G LTE Networks

### IMT-Advanced Requirements

An IMT-Advanced cellular system must fulfill the following requirements:

- Be based on an all-IP packet switched network.

- Have peak data rates of up to approximately 100 Mbit/s for high mobility such as mobile access and up to approximately 1 Gbit/s for low mobility such as no-madic/local wireless access.

- Be able to dynamically share and use the network resources to support more simultaneous users per cell.

- Use scalable channel bandwidths of 5–20 MHz, optionally up to 40 MHz.

- Have peak link spectral efficiency of 15 bit/s·Hz in the downlink, and 6.75 bit/s·Hz in the uplink (meaning that 1 Gbit/s in the downlink should be possible over less than 67 MHz bandwidth).

- System spectral efficiency is, in indoor cases, 3 bit/s·Hz·cell for downlink and 2.25 bit/s·Hz·cell for uplink.

- Smooth handovers across heterogeneous networks.

In September 2009, the technology proposals were submitted to the International Telecommunication Union (ITU) as 4G candidates. Basically all proposals are based on two technologies:

- LTE Advanced standardized by the 3GPP.

- 802.16m standardized by the IEEE.

Implementations of Mobile WiMAX and first-release LTE were largely considered a stopgap solution that would offer a considerable boost until WiMAX 2 (based on the 802.16m specification) and LTE Advanced was deployed. The latter's standard versions were ratified in spring 2011.

The first set of 3GPP requirements on LTE Advanced was approved in June 2008. LTE Advanced was standardized in 2010 as part of Release 10 of the 3GPP specification.

Some sources consider first-release LTE and Mobile WiMAX implementations as pre-4G or near-4G, as they do not fully comply with the planned requirements of 1 Gbit/s for stationary reception and 100 Mbit/s for mobile.

Confusion has been caused by some mobile carriers who have launched products advertised as 4G but which according to some sources are pre-4G versions, commonly referred to as 3.9G, which do not follow the ITU-R defined principles for 4G standards, but today can be called 4G according to ITU-R. Vodafone Netherlands for example, advertised LTE as 4G, while advertising LTE Advanced as their '4G+' service. A common argument for branding 3.9G systems as new-generation is that they use different frequency bands from 3G technologies; that they are based on a new radio-interface paradigm; and that the standards are not backwards compatible with 3G, whilst some of the standards are forwards compatible with IMT-2000 compliant versions of the same standards.

## Principal Technologies in all Candidate Systems

### Key Features

The following key features can be observed in all suggested 4G technologies:

- Physical layer transmission techniques are as follows:

  - MIMO: To attain ultra-high spectral efficiency by means of spatial processing including multi-antenna and multi-user MIMO.

  - Frequency-domain-equalization, for example multi-carrier modulation (OFDM) in the downlink or single-carrier frequency-domain-equalization (SC-FDE) in the uplink: To exploit the frequency selective channel property without complex equalization.

  - Frequency-domain statistical multiplexing, for example (OFDMA) or (single-carrier FDMA) (SC-FDMA, a.k.a. linearly precoded OFDMA, LP-OFDMA) in the uplink: Variable bit rate by assigning different sub-channels to different users based on the channel conditions.

  - Turbo principle error-correcting codes: To minimize the required SNR at the reception side.

- Channel-dependent scheduling: To use the time-varying channel.

- Link adaptation: Adaptive modulation and error-correcting codes.

- Mobile IP utilized for mobility.

- IP-based femtocells (home nodes connected to fixed Internet broadband infrastructure).

As opposed to earlier generations, 4G systems do not support circuit switched telephony. IEEE 802.20, UMB and OFDM standards lack soft-handover support, also known as cooperative relaying.

### Multiplexing and Access Schemes

Recently, new access schemes like Orthogonal FDMA (OFDMA), Single Carrier FDMA (SC-FDMA), Interleaved FDMA, and Multi-carrier CDMA (MC-CDMA) are gaining more importance for the next generation systems. These are based on efficient FFT algorithms and frequency domain equalization, resulting in a lower number of multiplications per second. They also make it possible to control the bandwidth and form the spectrum in a flexible way. However, they require advanced dynamic channel allocation and adaptive traffic scheduling.

WiMax is using OFDMA in the downlink and in the uplink. For the LTE (telecommunication), OFDMA is used for the downlink; by contrast, Single-carrier FDMA is used for the uplink since OFDMA contributes more to the PAPR related issues and results in nonlinear operation of amplifiers. IFDMA provides less power fluctuation and thus requires energy-inefficient linear amplifiers. Similarly, MC-CDMA is in the proposal for the IEEE 802.20 standard. These access schemes offer the same efficiencies as older technologies like CDMA. Apart from this, scalability and higher data rates can be achieved.

The other important advantage of the above-mentioned access techniques is that they require less complexity for equalization at the receiver. This is an added advantage especially in the MIMO environments since the spatial multiplexing transmission of MIMO systems inherently require high complexity equalization at the receiver.

In addition to improvements in these multiplexing systems, improved modulation techniques are being used. Whereas earlier standards largely used Phase-shift keying, more efficient systems such as 64QAM are being proposed for use with the 3GPP Long Term Evolution standards.

## IPv6 Support

Unlike 3G, which is based on two parallel infrastructures consisting of circuit switched and packet switched network nodes, 4G is based on packet switching only. This requires low-latency data transmission.

As IPv4 addresses are (nearly) exhausted, IPv6 is essential to support the large number of wireless-enabled devices that communicate using IP. By increasing the number of IP addresses available, IPv6 removes the need for network address translation (NAT), a method of sharing a limited number of addresses among a larger group of devices, which has a number of problems and limitations. When using IPv6, some kind of NAT is still required for communication with legacy IPv4 devices that are not also IPv6-connected.

As of June 2009, Verizon has posted Specifications that require any 4G devices on its network to support IPv6.

## Advanced Antenna Systems

The performance of radio communications depends on an antenna system, termed smart or intelligent antenna. Recently, multiple antenna technologies are emerging to achieve the goal of 4G systems such as high rate, high reliability, and long range communications. In the early 1990s, to cater for the growing data rate needs of data communication, many transmission schemes were proposed. One technology, spatial multiplexing, gained importance for its bandwidth conservation and power efficiency. Spatial multiplexing involves deploying multiple antennas at the transmitter and at the receiver. Independent streams can then be transmitted simultaneously from all the antennas. This technology, called MIMO (as a branch of intelligent antenna), multiplies the base data rate by (the smaller of) the number of transmit antennas or the number of receive antennas. Apart from this, the reliability in transmitting high speed data in the fading channel can be improved by using more antennas at the transmitter or at the receiver. This is called transmit or receive diversity. Both transmit/receive diversity and transmit spatial multiplexing are categorized into the space-time coding techniques, which does not necessarily require the channel knowledge at the transmitter. The other category is closed-loop multiple antenna technologies, which require channel knowledge at the transmitter.

## Open-wireless Architecture and Software-defined Radio

One of the key technologies for 4G and beyond is called Open Wireless Architecture (OWA), supporting multiple wireless air interfaces in an open architecture platform.

SDR is one form of open wireless architecture (OWA). Since 4G is a collection of wireless standards, the final form of a 4G device will constitute various standards. This can be efficiently realized using SDR technology, which is categorized to the area of the radio convergence.

## Disadvantages

4G introduces a potential inconvenience for those who travel internationally or wish to switch carriers. In order to make and receive 4G voice calls, the subscriber handset must not only have a matching frequency band (and in some cases require unlocking), it must also have the matching enablement settings for the local carrier and/or country. While a phone purchased from a given carrier can be expected to work with that carrier, making 4G voice calls on another carrier's network (including international roaming) may be impossible without a software update specific to the local carrier and the phone model in question, which may or may not be available (although fallback to 3G for voice calling may still be possible if a 3G network is available with a matching frequency band).

## Beyond 4G Research

A major issue in 4G systems is to make the high bit rates available in a larger portion of the cell, especially to users in an exposed position in between several base stations. In current research, this issue is addressed by macro-diversity techniques, also known as group cooperative relay, and also by Beam-Division Multiple Access (BDMA).

Pervasive networks are an amorphous and at present entirely hypothetical concept where the user can be simultaneously connected to several wireless access technologies and can seamlessly move between them. These access technologies can be Wi-Fi, UMTS, EDGE, or any other future access technology. Included in this concept is also smart-radio (also known as cognitive radio) technology to efficiently manage spectrum use and transmission power as well as the use of mesh routing protocols to create a pervasive network.

## Fifth Generation (5G)

Fifth-generation wireless (5G) is the latest iteration of cellular technology, engineered to greatly increase the speed and responsiveness of wireless networks. With 5G, data transmitted over wireless broadband connections can travel at multigigabit speeds, with potential peak speeds as high as 20 gigabits per second (Gbps) by some estimates. These speeds exceed wireline network speeds and offer latency of 1 millisecond (ms) or lower for uses that require real-time feedback. 5G will also enable a sharp increase in the amount of data transmitted over wireless systems due to more available bandwidth and advanced antenna technology.

Fifth-generation wireless (5G) is the latest iteration of cellular technology, engineered to greatly increase the speed and responsiveness of wireless networks. With 5G, data transmitted over wireless broadband connections can travel at multigigabit speeds, with potential peak speeds as high as 20 gigabits per second (Gbps) by some estimates. These speeds exceed wireline network speeds and offer latency of 1 millisecond (ms) or lower for uses that require real-time feedback. 5G will also enable a sharp increase in the amount of data transmitted over wireless systems due to more available bandwidth and advanced antenna technology.

### Working of 5G

Wireless networks are composed of cell sites divided into sectors that send data through radio waves. Fourth-generation (4G) Long-Term Evolution (LTE) wireless technology provides the foundation for 5G. Unlike 4G, which requires large, high-power cell towers to radiate signals over longer distances, 5G wireless signals will be transmitted via large numbers of small cell stations located in places like light poles or building roofs. The use of multiple small cells is necessary because the millimeter wave (MM wave) spectrum the band of spectrum between 30 gigahertz and 300 GHz that 5G relies on to generate high speeds can only travel over short distances and is subject to interference from weather and physical obstacles, like buildings or trees.

Previous generations of wireless technology have used lower-frequency bands of spectrum. To offset the challenges relating to distance and interference with MM waves, the wireless industry is also considering the use of a lower-frequency spectrum for 5G networks so network operators could use spectrum they already own to build out their new networks. Lower-frequency spectrum reaches greater distances but has lower speed and capacity than MM wave.

The lower frequency wireless spectrum is made up of low- and midband frequencies. Low-band frequencies operate at around 600 to 700 megahertz (MHz), while midband frequencies operate at around 2.5 to 3.5 GHz. This is compared to high-band MM wave signals, which operate at approximately 24 to 39 GHz.

MM wave signals can be easily blocked by objects such as trees, walls and buildings meaning that, much of the time, MM waves can only cover about a city block within direct line of sight of a cell site or node. Different approaches have been tackled regarding how to get around this issue. A brute-force approach involves using multiple nodes around each block of a populated area so that a 5G-enabled device can use an air interface switching from node to node while maintaining MM wave speeds.

Another approach the more feasible one for creating a national 5G network is to use a combination of high-, medium- and low-band frequencies. MM wave may be used in densely populated areas, while low- and midband nodes may be used in less dense areas. The low-band frequencies can travel longer and through different objects. One low-band 5G node can stay connected to a 5G-enabled device for up to hundreds of square miles. This means that an implementation of all three bands will give blanketed coverage while providing the fastest speeds in the most highly trafficked areas.

## Speed of 5G Networks

5G download speeds can currently reach upwards of 1,000 megabits per second (Mbps) or even up to 2.1 Gbps. To visualize this, a user could start a YouTube video in 1080p quality on a 5G device without it buffering. Downloading an app or an episode of a Netflix show, which may currently take up to a few minutes, can be completed in just a few seconds. Wirelessly streaming video in 4K also becomes much more viable. If on MM wave, these examples would currently need to be within an unobstructed city block away from a 5G node; if not, the download speed would drop back down to 4G.

Low band can stay locked at 5G over longer distances, and even though the overall speed of low-band 5G may be slower than MM wave, low band should still be faster than what would be considered a good 4G connection. Low-band 5G download speeds may be up to 30 to 250 Mbps. Low-band 5G is more likely to be available for more rural locations. Midband 5G download speeds may reach up to 100 to 900 Mbps, and it is likely to be used in major metro areas.

## Benefits of 5G

Even though the downsides of 5G are clear when considering how easily MM waves can be blocked, or less clear considering radio frequency (RF) exposure limits, 5G still has plenty of worthy benefits, such as the following:

- Use of higher frequencies.

- High bandwidth.

- Enhanced mobile broadband.

- A lower latency of 1 ms.

- Higher data rates, which will enable new technology options over 5G networks, such as 4K streaming or near-real-time streaming of virtual reality (VR).

- The potential to have a 5G mobile network made up of low-band, midband and MM wave frequencies.

## References

- Mobile-telephone, technology: britannica.com, Retrieved 26 February, 2020

- Bbc-the-spectrum-and-its-uses, spectrum: downloads.bbc.co.uk, Retrieved 21 August, 2020

- Wifi-networking-radio-wave-basics, wireless-infrastructure: networkcomputing.com, Retrieved 20 July, 2020

- Cell-phones-work: pongcase.com, Retrieved 25 January, 2020

- Communication-using-infrared-technology: elprocus.com, Retrieved 19 June, 2020

- Introduction-to-mobile-satellite-communication-system-and-its-services: efxkits.us, Retrieved 20 May, 2018

- CDMA-Code-Division-Multiple-Access: searchnetworking.techtarget.com, Retrieved 26 July, 2020

# Wireless Networks

- **Wireless Application Protocol**
- **WAP Gateway**

Wireless networks do not make use of wires and cables for establishing communication. It includes Bluetooth, hiperLAN and wireless PAN, LAN, MAN, WAN, and other wireless applications. The topics elaborated in this chapter will help in gaining a better perspective of the subject of wireless networks.

Wireless network is a network set up by using radio signal frequency to communicate among computers and other network devices. Sometimes it's also referred to as WiFi network or WLAN. This network is getting popular nowadays due to easy to setup feature and no cabling involved. You can connect computers anywhere in your home without the need for wires.

Here is simple explanation of how it works, let say you have 2 computers each equipped with wireless adapter and you have set up wireless router. When the computer send out the data, the binary data will be encoded to radio frequency and transmitted via wireless router. The receiving computer will then decode the signal back to binary data.

It doesn't matter you are using broadband cable/DSL modem to access internet, both ways will work with wireless network. If you heard about wireless hotspot, that means that location is equipped with wireless devices for you and others to join the network.

The two main components are wireless router or access point and wireless clients.

If you have not set up any wired network, then just get a wireless routerand attach it to cable or DSL modem. You then set up wireless client by adding wireless card to each computer and form a simple wireless network. You can also cable connect computer directly to router if there are switch ports available.

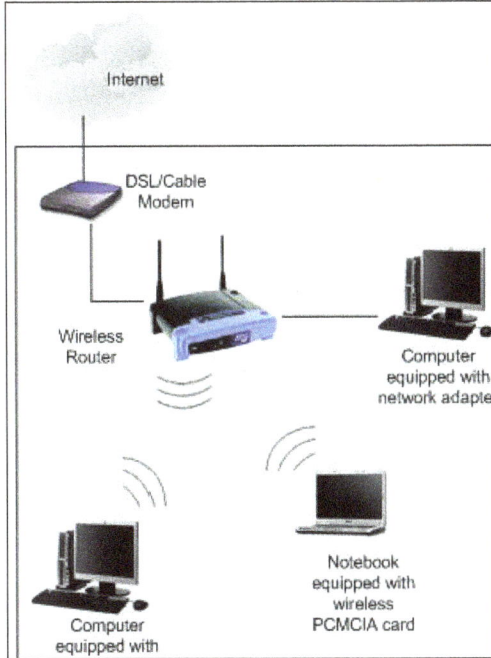

If you already have wired Ethernet network at home, you can attach a wireless access point to existing network router and have wireless access at home.

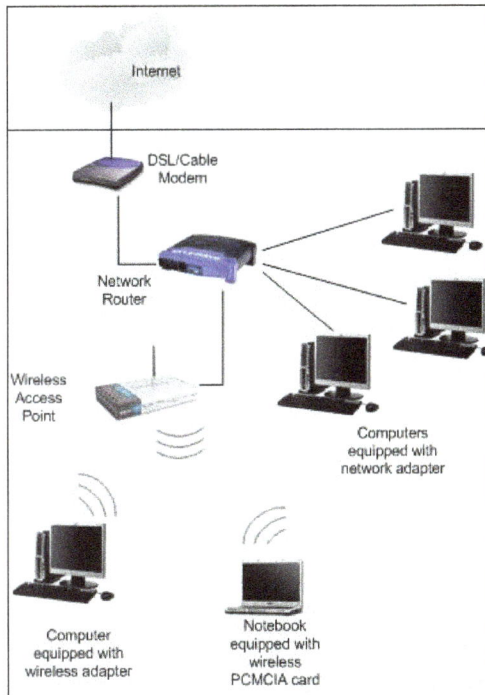

Wireless router or access points should be installed in a way that maximizes coverage as well as throughput. The coverage provided is generally referred to as the coverage cell. Large areas usually require more than one access point in order to have adequate coverage. You can also add access point to your existing wireless router to improve coverage.

## Wireless Operating Mode

The IEEE 802.11 standards specify two operating modes: infrastructure mode and ad hoc mode.

Infrastructure mode is used to connect computers with wireless network adapters, also known as wireless clients, to an existing wired network with the help from wireless router or access point.

Ad hoc mode is used to connect wireless clients directly together, without the need for a wireless router or access point. An ad hoc network consists of up to 9 wireless clients, which send their data directly to each other.

## Bluetooth

Bluetooth is a wireless technology standard used for exchanging data between fixed and mobile devices over short distances using short-wavelength UHF radio waves in the industrial, scientific and medical radio bands, from 2.400 to 2.485 GHz, and building personal area networks (PANs). It was originally conceived as a wireless alternative to RS-232 data cables.

Bluetooth is managed by the Bluetooth Special Interest Group (SIG), which has more than 35,000 member companies in the areas of telecommunication, computing, networking, and consumer electronics. The IEEE standardized Bluetooth as IEEE 802.15.1, but no longer maintains the standard. The Bluetooth SIG oversees development of the specification, manages the qualification program, and protects the trademarks. A manufacturer must meet Bluetooth SIG standards to market it as a Bluetooth device. A network of patents apply to the technology, which are licensed to individual qualifying devices. As of 2009, Bluetooth integrated circuit chips ship approximately 920 million units annually.

## Implementation

Bluetooth operates at frequencies between 2.402 and 2.480 GHz, or 2.400 and 2.4835 GHz including guard bands 2 MHz wide at the bottom end and 3.5 MHz wide at the top. This is in the globally unlicensed (but not unregulated) industrial, scientific and medical (ISM) 2.4 GHz short-range radio frequency band. Bluetooth uses a radio technology called frequency-hopping spread spectrum. Bluetooth divides transmitted data into packets, and transmits each packet on one of 79 designated Bluetooth channels. Each channel has a bandwidth of 1 MHz. It usually performs 1600 hops per second, with adaptive frequency-hopping (AFH) enabled. Bluetooth Low Energy uses 2 MHz spacing, which accommodates 40 channels.

Originally, Gaussian frequency-shift keying (GFSK) modulation was the only modulation scheme available. Since the introduction of Bluetooth 2.0+EDR, $\pi/4$-DQPSK (differential quadrature phase-shift keying) and 8-DPSK modulation may also be used between compatible devices. Devices functioning with GFSK are said to be operating in basic rate (BR) mode where an instantaneous bit rate of 1 Mbit/s is possible. The term Enhanced Data Rate (EDR) is used to describe $\pi/4$-DPSK and 8-DPSK schemes, each giving 2 and 3 Mbit/s respectively. The combination of these (BR and EDR) modes in Bluetooth radio technology is classified as a BR/EDR radio.

Bluetooth is a packet-based protocol with master/slave architecture. One master may communicate with up to seven slaves in a piconet. All devices share the master's clock. Packet exchange is based on the basic clock, defined by the master, which ticks at 312.5 μs intervals. Two clock ticks make up a slot of 625 μs, and two slots make up a slot pair of 1250 μs. In the simple case of single-slot packets, the master transmits in even slots and receives in odd slots. The slave, conversely, receives in even slots and transmits in odd slots. Packets may be 1, 3 or 5 slots long, but in all cases the master's transmission begins in even slots and the slave's in odd slots.

The above excludes Bluetooth Low Energy, introduced in the 4.0 specification, which uses the same spectrum but somewhat differently.

## Communication and Connection

A master BR/EDR Bluetooth device can communicate with a maximum of seven devices in a piconet (an ad-hoc computer network using Bluetooth technology), though not all devices reach this maximum. The devices can switch roles, by agreement, and the slave can become the master (for example, a headset initiating a connection to a phone necessarily begins as master as an initiator of the connection but may subsequently operate as the slave).

The Bluetooth Core Specification provides for the connection of two or more piconets to form a scatternet, in which certain devices simultaneously play the master role in one piconet and the slave role in another.

At any given time, data can be transferred between the master and one other device (except for the little-used broadcast mode). The master chooses which slave device to address; typically, it switches rapidly from one device to another in a round-robin fashion. Since it is the master that chooses which slave to address, whereas a slave is (in theory) supposed to listen in each receive slot, being a master is a lighter burden than being a slave. Being a master of seven slaves is possible; being a slave of more than one master is possible. The specification is vague as to required behavior in scatternets.

## Uses of Communication and Connection

Bluetooth is a standard wire-replacement communications protocol primarily designed for low power consumption, with a short range based on low-cost transceiver micro-chips in each device. Because the devices use a radio (broadcast) communications system, they do not have to be in visual line of sight of each other; however, a quasi-optical wireless path must be viable. Range is power-class-dependent, but effective ranges vary in practice.

Officially Class 3 radios have a range of up to 1 metre (3 ft.), Class 2, most commonly found in mobile devices, 10 metres (33 ft.), and Class 1, primarily for industrial use cases,100 metres (300 ft.). Bluetooth Marketing qualifies that Class 1 range is in most cases 20–30 metres (66–98 ft.) and Class 2 range 5–10 metres (16–33 ft.). The actual range achieved by a given link will depend on the qualities of the devices at both ends of the link, as well as the air conditions in between, and other factors.

The effective range varies depending on propagation conditions, material coverage, production sample variations, antenna configurations and battery conditions. Most Bluetooth applications are for indoor conditions, where attenuation of walls and signal fading due to signal reflections make the range far lower than specified line-of-sight ranges of the Bluetooth products.

Most Bluetooth applications are battery-powered Class 2 devices, with little difference in range whether the other end of the link is a Class 1 or Class 2 device as the lower-powered device tends to set the range limit. In some cases the effective range of the data link can be extended when a Class 2 device is connecting to a Class 1 transceiver with both higher sensitivity and transmission power than a typical Class 2 device. Mostly, however, the Class 1 devices have a similar sensitivity to Class 2 devices. Connecting two Class 1 devices with both high sensitivity and high power can allow ranges far in excess of the typical 100m, depending on the throughput required by the application. Some such devices allow open field ranges of up to 1 km and beyond between two similar devices without exceeding legal emission limits.

The Bluetooth Core Specification mandates a range of not less than 10 metres (33 ft.), but there is no upper limit on actual range. Manufacturers' implementations can be tuned to provide the range needed for each case.

| Ranges of Bluetooth devices by class | | | |
|---|---|---|---|
| Class | Max. permitted power | | Typ. range (m) |
| | (mW) | (dBm) | |
| 1 | 100 | 20 | ~100 |
| 1.5 (BT 5 Vol 6 Part A Sect 3) | 10 | 10 | ~20 |
| 2 | 2.5 | 4 | ~10 |
| 3 | 1 | 0 | ~1 |
| 4 | 0.5 | −3 | ~0.5 |

## Bluetooth Profile

To use Bluetooth wireless technology, a device must be able to interpret certain

Bluetooth profiles, which are definitions of possible applications and specify general behaviors that Bluetooth-enabled devices use to communicate with other Bluetooth devices. These profiles include settings to parameterize and to control the communication from the start. Adherence to profiles saves the time for transmitting the parameters anew before the bi-directional link becomes effective. There are a wide range of Bluetooth profiles that describe many different types of applications or use cases for devices.

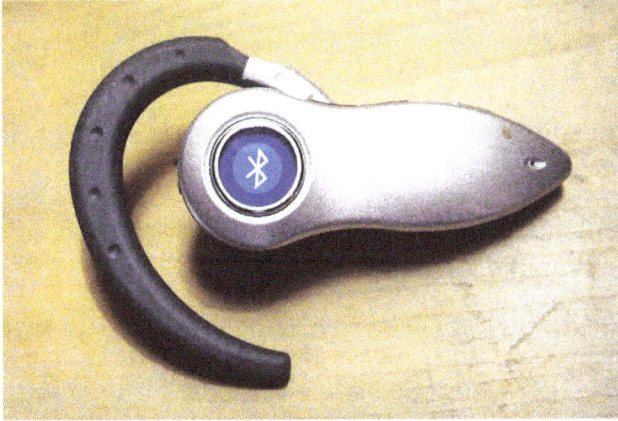

A typical Bluetooth mobile phone headset.

## List of Applications

- Wireless control and communication between a mobile phone and a handsfree headset. This was one of the earliest applications to become popular.

- Wireless control of and communication between a mobile phone and a Bluetooth compatible car stereo system.

- Wireless communication between a smartphone and a smart lock for unlocking doors.

- Wireless control of and communication with iOS and Android device phones, tablets and portable wireless speakers.

- Wireless Bluetooth headset and Intercom. Idiomatically, a headset is sometimes called "a Bluetooth".

- Wireless streaming of audio to headphones with or without communication capabilities.

- Wireless streaming of data collected by Bluetooth-enabled fitness devices to phone or PC.

- Wireless networking between PCs in a confined space and where little bandwidth is required.

- Wireless communication with PC input and output devices, the most common being the mouse, keyboard and printer.

- Transfer of files, contact details, calendar appointments, and reminders between devices with OBEX.

- Replacement of previous wired RS-232 serial communications in test equipment, GPS receivers, medical equipment, bar code scanners, and traffic control devices.

- For controls where infrared was often used.

- For low bandwidth applications where higher USB bandwidth is not required and cable-free connection desired.

- Sending small advertisements from Bluetooth-enabled advertising hoardings to other, discoverable, Bluetooth devices.

- Wireless bridge between two Industrial Ethernet (e.g., PROFINET) networks.

- Seventh and eighth generation game consoles such as Nintendo's Wii, and Sony's PlayStation 3 use Bluetooth for their respective wireless controllers.

- Dial-up internet access on personal computers or PDAs using a data-capable mobile phone as a wireless modem.

- Short-range transmission of health sensor data from medical devices to mobile phone, set-top box or dedicated telehealth devices.

- Allowing a DECT phone to ring and answer calls on behalf of a nearby mobile phone.

- Real-time location systems (RTLS) are used to track and identify the location of objects in real time using "Nodes" or "tags" attached to, or embedded in, the objects tracked, and "Readers" that receive and process the wireless signals from these tags to determine their locations.

- Personal security application on mobile phones for prevention of theft or loss of items. The protected item has a Bluetooth marker (e.g., a tag) that is in constant communication with the phone. If the connection is broken (the marker is out of range of the phone) then an alarm is raised. This can also be used as a man overboard alarm. A product using this technology has been available since 2009.

- Calgary, Alberta, Canada's Roads Traffic division uses data collected from travelers' Bluetooth devices to predict travel times and road congestion for motorists.

- Wireless transmission of audio (a more reliable alternative to FM transmitters).

- Live video streaming to the visual cortical implant device by Nabeel Fattah in Newcastle University 2017.

- Connection of motion controllers to a PC when using VR headsets.

## Bluetooth vs. Wi-Fi (IEEE 802.11)

Bluetooth and Wi-Fi (Wi-Fi is the brand name for products using IEEE 802.11 standards) have some similar applications: setting up networks, printing, or transferring files. Wi-Fi is intended as a replacement for high-speed cabling for general local area network access in work areas or home. This category of applications is sometimes called wireless local area networks (WLAN). Bluetooth was intended for portable equipment and its applications. The category of applications is outlined as the wireless personal area network (WPAN). Bluetooth is a replacement for cabling in a variety of personally carried applications in any setting, and also works for fixed location applications such as smart energy functionality in the home (thermostats, etc.).

Wi-Fi and Bluetooth are to some extent complementary in their applications and usage. Wi-Fi is usually access point-centered, with an asymmetrical client-server connection with all traffic routed through the access point, while Bluetooth is usually symmetrical, between two Bluetooth devices. Bluetooth serves well in simple applications where two devices need to connect with a minimal configuration like a button press, as in headsets and remote controls, while Wi-Fi suits better in applications where some degree of client configuration is possible and high speeds are required, especially for network access through an access node. However, Bluetooth access points do exist, and ad-hoc connections are possible with Wi-Fi though not as simply as with Bluetooth. Wi-Fi Direct was recently developed to add more Bluetooth-like ad-hoc functionality to Wi-Fi.

## Devices

A Bluetooth USB dongle with a 100 m range.

Bluetooth exists in numerous products such as telephones, speakers, tablets, media players, robotics systems, laptops, and console gaming equipment as well as some high definition headsets, modems, hearing aids and even watches. Given the variety of devices which use the Bluetooth, coupled with the contemporary deprecation of headphone jacks by Apple, Google, and other companies, and the lack of regulation by the FCC, the technology is prone to interference. Nonetheless Bluetooth is useful when transferring information between two or more devices that are near each other in low-bandwidth situations. Bluetooth is commonly used to transfer sound data with telephones (i.e., with a Bluetooth headset) or byte data with hand-held computers (transferring files).

Bluetooth protocols simplify the discovery and setup of services between devices. Bluetooth devices can advertise all of the services they provide. This makes using services easier, because more of the security, network address and permission configuration can be automated than with many other network types.

A typical Bluetooth USB dongle.

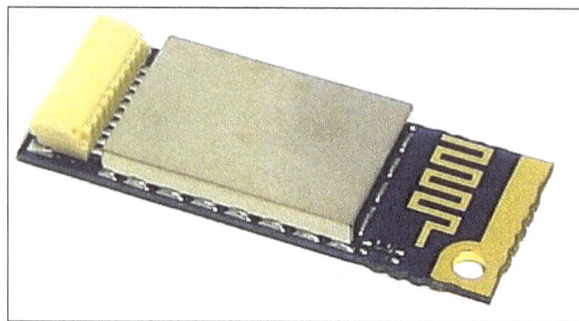

An internal notebook Bluetooth card (14×36×4 mm).

## Computer Requirements

A personal computer that does not have embedded Bluetooth can use a Bluetooth adapter that enables the PC to communicate with Bluetooth devices. While some desktop computers and most recent laptops come with a built-in Bluetooth radio, others require an external adapter, typically in the form of a small USB "dongle".

Unlike its predecessor, IrDA, which requires a separate adapter for each device, Bluetooth lets multiple devices communicate with a computer over a single adapter.

## Operating System Implementation

For Microsoft platforms, Windows XP Service Pack 2 and SP3 releases work natively with Bluetooth v1.1, v2.0 and v2.0+EDR. Previous versions required users to install their Bluetooth adapter's own drivers, which were not directly supported by Microsoft. Microsoft's own Bluetooth dongles (packaged with their Bluetooth computer devices) have no external drivers and thus require at least Windows XP Service Pack 2. Windows Vista RTM/SP1 with the Feature Pack for Wireless or Windows Vista SP2 work with Bluetooth v2.1+EDR. Windows 7 works with Bluetooth v2.1+EDR and Extended Inquiry Response (EIR). The Windows XP and Windows Vista/Windows 7 Bluetooth stacks support the following Bluetooth profiles natively: PAN, SPP, DUN, HID, HCRP. The Windows XP stack can be replaced by a third party stack that supports more profiles or newer Bluetooth versions. The Windows Vista/Windows 7 Bluetooth stack supports vendor-supplied additional profiles without requiring that the Microsoft stack be replaced. It is generally recommended to install the latest vendor driver and its associated stack to be able to use the Bluetooth device at its fullest extent.

Apple products have worked with Bluetooth since Mac OS X v10.2, which was released in 2002.

Linux has two popular Bluetooth stacks, BlueZ and Fluoride. The BlueZ stack is included with most Linux kernels and was originally developed by Qualcomm. Fluoride, earlier known as Bluedroid is included in Android OS and was originally developed by Broadcom. There is also Affix stack, developed by Nokia. It was once popular, but has not been updated since 2005.

FreeBSD has included Bluetooth since its v5.0 release, implemented through netgraph.

NetBSD has included Bluetooth since its v4.0 release. Its Bluetooth stack was ported to OpenBSD as well, however OpenBSD later removed it as unmaintained.

DragonFly BSD has had NetBSD's Bluetooth implementation since 1.11. A netgraph-based implementation from FreeBSD has also been available in the tree, possibly disabled until 2014-11-15, and may require more work.

## Specifications and Features

The specifications were formalized by the Bluetooth Special Interest Group (SIG) and formally announced on the 20 of May 1998. Today it has a membership of over 30,000 companies worldwide. It was established by Ericsson, IBM, Intel, Nokia and Toshiba, and later joined by many other companies.

All versions of the Bluetooth standards support downward compatibility. That lets the latest standard cover all older versions.

The Bluetooth Core Specification Working Group (CSWG) produces mainly 4 kinds of specifications:

- The Bluetooth Core Specification, release cycle is typically a few years in between.

- Core Specification Addendum (CSA), release cycle can be as tight as a few times per year.

- Core Specification Supplements (CSS), can be released very quickly.

- Errata (Available with a user account: Errata login.

## Bluetooth 1.0 and 1.0B

Versions 1.0 and 1.0B had many problems, and manufacturers had difficulty making their products interoperable. Versions 1.0 and 1.0B also included mandatory Bluetooth hardware device address (BD_ADDR) transmission in the Connecting process (rendering anonymity impossible at the protocol level), which was a major setback for certain services planned for use in Bluetooth environments.

## Bluetooth 1.1

- Ratified as IEEE Standard 802.15.1–2002.

- Many errors found in the v1.0B specifications were fixed.

- Added possibility of non-encrypted channels.

- Received Signal Strength Indicator (RSSI).

## Bluetooth 1.2

Major enhancements include:

- Faster Connection and Discovery.

- Adaptive frequency-hopping spread spectrum (AFH), which improves resistance to radio frequency interference by avoiding the use of crowded frequencies in the hopping sequence.

- Higher transmission speeds in practice than in v1.1, up to 721 Kbit/s.

- Extended Synchronous Connections (eSCO), which improve voice quality of audio links by allowing retransmissions of corrupted packets, and may optionally increase audio latency to provide better concurrent data transfer.

- Host Controller Interface (HCI) operation with three-wire UART.

- Ratified as IEEE Standard 802.15.1–2005.

- Introduced Flow Control and Retransmission Modes for L2CAP.

## Bluetooth 2.0 + EDR

This version of the Bluetooth Core Specification was released before 2005. The main difference is the introduction of an Enhanced Data Rate (EDR) for faster data transfer. The bit rate of EDR is 3 Mbit/s, although the maximum data transfer rate (allowing for inter-packet time and acknowledgements) is 2.1 Mbit/s. EDR uses a combination of GFSK and phase-shift keying modulation (PSK) with two variants, $\pi/4$-DQPSK and 8-DPSK. EDR can provide a lower power consumption through a reduced duty cycle.

The specification is published as Bluetooth v2.0 + EDR, which implies that EDR is an optional feature. Aside from EDR, the v2.0 specification contains other minor improvements, and products may claim compliance to "Bluetooth v2.0" without supporting the higher data rate. At least one commercial device states "Bluetooth v2.0 without EDR" on its data sheet.

## Bluetooth 2.1 + EDR

Bluetooth Core Specification Version 2.1 + EDR was adopted by the Bluetooth SIG on 26 July 2007.

The headline feature of v2.1 is secure simple pairing (SSP): this improves the pairing experience for Bluetooth devices, while increasing the use and strength of security.

Version 2.1 allows various other improvements, including extended inquiry response (EIR), which provides more information during the inquiry procedure to allow better filtering of devices before connection; and sniff subrating, which reduces the power consumption in low-power mode.

## Bluetooth 3.0 + HS

Version 3.0 + HS of the Bluetooth Core Specification were adopted by the Bluetooth SIG on 21 April 2009. Bluetooth v3.0 + HS provide theoretical data transfer speeds of up to 24 Mbit/s, though not over the Bluetooth link itself. Instead, the Bluetooth link is used for negotiation and establishment, and the high data rate traffic is carried over a colocated 802.11 link.

The main new feature is AMP (Alternative MAC/PHY), the addition of 802.11 as a high-speed transport. The high-speed part of the specification is not mandatory, and hence only devices that display the "+HS" logo actually support Bluetooth over 802.11 high-speed data transfer. A Bluetooth v3.0 device without the "+HS" suffix is only required to support features introduced in Core Specification Version 3.0 or earlier Core Specification Addendum 1.

## L2CAP Enhanced Modes

Enhanced Retransmission Mode (ERTM) implements reliable L2CAP channel, while Streaming Mode (SM) implements unreliable channel with no retransmission or flow control. Introduced in Core Specification Addendum 1.

## Alternative MAC/PHY

Enables the use of alternative MAC and PHYs for transporting Bluetooth profile data. The Bluetooth radio is still used for device discovery, initial connection and profile configuration. However, when large quantities of data must be sent, the high-speed alternative MAC PHY 802.11 (typically associated with Wi-Fi) transports the data. This means that Bluetooth uses proven low power connection models when the system is idle, and the faster radio when it must send large quantities of data. AMP links require enhanced L2CAP modes.

## Unicast Connectionless Data

Permits sending service data without establishing an explicit L2CAP channel. It is intended for use by applications that require low latency between user action and reconnection/transmission of data. This is only appropriate for small amounts of data.

## Enhanced Power Control

Updates the power control feature to remove the open loop power control, and also to clarify ambiguities in power control introduced by the new modulation schemes added for EDR. Enhanced power control removes the ambiguities by specifying the behaviour that is expected. The feature also adds closed loop power control, meaning RSSI filtering can start as the response is received. Additionally, a "go straight to maximum power" request has been introduced. This is expected to deal with the headset link loss issue typically observed when a user puts their phone into a pocket on the opposite side to the headset.

## Ultra-wideband

The high-speed (AMP) feature of Bluetooth v3.0 was originally intended for UWB, but the WiMedia Alliance, the body responsible for the flavor of UWB intended for Bluetooth, announced in March 2009 that it was disbanding, and ultimately UWB was omitted from the Core v3.0 specification.

On 16 March 2009, the WiMedia Alliance announced it was entering into technology transfer agreements for the WiMedia Ultra-wideband (UWB) specifications. WiMedia has transferred all current and future specifications, including work on future high-speed and power-optimized implementations, to the Bluetooth Special Interest Group (SIG), Wireless USB Promoter Group and the USB Implementers Forum. After successful completion of the technology transfer, marketing, and related administrative items, the WiMedia Alliance ceased operations.

In October 2009 the Bluetooth Special Interest Group suspended development of UWB as part of the alternative MAC/PHY, Bluetooth v3.0 + HS solution. A small, but significant, number of former WiMedia members had not and would not sign up to the necessary agreements for the IP transfer. The Bluetooth SIG is now in the process of evaluating other options for its longer term roadmap.

## Bluetooth 4.0

The Bluetooth SIG completed the Bluetooth Core Specification version 4.0 (called Bluetooth Smart) and has been adopted as of 30 June 2010. It includes Classic Bluetooth, Bluetooth high speed and Bluetooth Low Energy (BLE) protocols. Bluetooth high speed is based on Wi-Fi, and Classic Bluetooth consists of legacy Bluetooth protocols.

Bluetooth Low Energy, previously known as Wibree, is a subset of Bluetooth v4.0 with an entirely new protocol stack for rapid build-up of simple links. As an alternative to the Bluetooth standard protocols that were introduced in Bluetooth v1.0 to v3.0, it is aimed at very low power applications powered by a coin cell. Chip designs allow for two types of implementation, dual-mode, single-mode and enhanced past versions. The provisional names Wibree and Bluetooth ULP (Ultra Low Power) were abandoned and the BLE name was used for a while. In late 2011, new logos "Bluetooth Smart Ready" for hosts and "Bluetooth Smart" for sensors were introduced as the general-public face of BLE.

Compared to Classic Bluetooth, Bluetooth Low Energy is intended to provide considerably reduced power consumption and cost while maintaining a similar communication range. In terms of lengthening the battery life of Bluetooth devices, BLE represents a significant progression.

- In a single-mode implementation, only the low energy protocol stack is implemented. Dialog Semiconductor, STMicroelectronics, AMICCOM, CSR, Nordic Semiconductor and Texas Instruments have released single mode Bluetooth Low Energy solutions.

- In a dual-mode implementation, Bluetooth Smart functionality is integrated into an existing Classic Bluetooth controller. As of March 2011, the following semiconductor companies have announced the availability of chips meeting the standard: Qualcomm-Atheros, CSR, Broadcom and Texas Instruments. The compliant architecture shares all of Classic Bluetooth's existing radio and functionality resulting in a negligible cost increase compared to Classic Bluetooth.

Cost-reduced single-mode chips, which enable highly integrated and compact devices, feature a lightweight Link Layer providing ultra-low power idle mode operation, simple device discovery, and reliable point-to-multipoint data transfer with advanced power-save and secure encrypted connections at the lowest possible cost.

General improvements in version 4.0 include the changes necessary to facilitate BLE modes, as well the Generic Attribute Profile (GATT) and Security Manager (SM) services with AES Encryption.

Core Specification Addendum 2 was unveiled in December 2011; it contains improvements to the audio Host Controller Interface and to the High Speed (802.11) Protocol Adaptation Layer.

Core Specification Addendum 3 revision 2 has an adoption date of 24 July 2012.

Core Specification Addendum 4 has an adoption date of 12 February 2013.

## Bluetooth 4.1

The Bluetooth SIG announced formal adoption of the Bluetooth v4.1 specification on 4 December 2013. This specification is an incremental software update to Bluetooth Specification v4.0, and not a hardware update. The update incorporates Bluetooth Core Specification Addenda (CSA 1, 2, 3 & 4) and adds new features that improve consumer usability. These include increased co-existence support for LTE, bulk data exchange rates and aid developer innovation by allowing devices to support multiple roles simultaneously.

New features of this specification include:

- Mobile Wireless Service Coexistence Signaling.

- Train Nudging and Generalized Interlaced Scanning.

- Low Duty Cycle Directed Advertising.

- L2CAP Connection Oriented and Dedicated Channels with Credit-Based Flow Control.

- Dual Mode and Topology.

- LE Link Layer Topology.

- 802.11n PAL.

- Audio Architecture Updates for Wide Band Speech.

- Fast Data Advertising Interval.

- Limited Discovery Time.

Notice that some features were already available in a Core Specification Addendum (CSA) before the release of v4.1.

## Bluetooth 4.2

Released on December 2, 2014, it introduces features for the Internet of Things.

The major areas of improvement are:

- Low Energy Secure Connection with Data Packet Length Extension.

- Link Layer Privacy with Extended Scanner Filter Policies.

- Internet Protocol Support Profile (IPSP) version 6 ready for Bluetooth Smart things to support connected home.

Older Bluetooth hardware may receive 4.2 features such as Data Packet Length Extension and improved privacy via firmware updates.

## Bluetooth 5

The Bluetooth SIG released Bluetooth 5 on December 6, 2016. Its new features are mainly focused on new Internet of Things technology. Sony was the first to announce Bluetooth 5.0 support with its Xperia XZ Premium in Feb 2017 during the Mobile World Congress 2017. The Samsung Galaxy S8 launched with Bluetooth 5 support in April 2017. In September 2017, the iPhone 8, 8 Plus and iPhone X launched with Bluetooth 5 support as well. Apple also integrated Bluetooth 5 in its new HomePod offering released on February 9, 2018. Marketing drops the point number; so that it is just "Bluetooth 5" (unlike Bluetooth 4.0). The change is for the sake of "Simplifying our marketing, communicating user benefits more effectively and making it easier to signal significant technology updates to the market".

Bluetooth 5 provides, for BLE, options that can double the speed (2 Mbit/s burst) at the expense of range, or up to fourfold the range at the expense of data rate. The increase in transmissions could be important for Internet of Things devices, where many nodes connect throughout a whole house. Bluetooth 5 adds functionality for connectionless services such as location-relevant navigation of low-energy Bluetooth connections.

The major areas of improvement are:

- Slot Availability Mask (SAM).

- 2 Mbit/s PHY for LE.

- LE Long Range.

- High Duty Cycle Non-Connectable Advertising.

- LE Advertising Extensions.

- LE Channel Selection Algorithm #2.

Features Added in CSA5 – Integrated in v5.0:

- Higher Output Power.

The following features were removed in this version of the specification:

- Park State.

## Bluetooth 5.1

The Bluetooth SIG presented Bluetooth 5.1 on 21 January 2019.

The major areas of improvement are:

- Angle of Arrival (AoA) and Angle of Departure (AoD) which are used for location and tracking of devices.

- Advertising Channel Index.

- GATT Caching.

Minor Enhancements batch 1:

- HCI support for debug keys in LE Secure Connections.

- Sleep clock accuracy update mechanism.

- ADI field in scan response data.

- Interaction between QoS and Flow Specification.

- Host channel classification for secondary advertising.

- Allow the SID to appear in scan response reports.

- Specify the behavior when rules are violated.

- Periodic Advertising Sync Transfer.

Features Added in Core Specification Addendum (CSA) 6 – Integrated in v5.1:

- Models.

- Mesh-based model hierarchy.

The following features were removed in this version of the specification:

- Unit keys.

## Bluetooth 5.2

On January 6, 2020, the Bluetooth SIG published the Bluetooth Core Specification Version 5.2. The new specification adds three main new features:

- Enhanced Attribute Protocol (EATT), an improved version of the Attribute Protocol (ATT).

- LE Power Control.

- LE Isochronous Channels.

## Technical Information

### Software

Seeking to extend the compatibility of Bluetooth devices, the devices that adhere to the standard use an interface called HCI (Host Controller Interface) between the host device (laptop, phone, etc.) and the Bluetooth device as such (Bluetooth chip).

High-level protocols such as the SDP (Protocol used to find other Bluetooth devices within the communication range, also responsible for detecting the function of devices in range), RFCOMM (Protocol used to emulate serial port connections) and TCS (Telephony control protocol) interact with the baseband controller through the L2CAP Protocol (Logical Link Control and Adaptation Protocol). The L2CAP protocol is responsible for the segmentation and reassembly of the packets.

### Hardware

The hardware that makes up the Bluetooth device is made up of, logically, two parts; which may or may not be physically separate. A radio device, responsible for modulating and transmitting the signal; and a digital controller. The digital controller is likely a CPU, one of whose functions is to run a Link Controller; and interfaces with the host device; but some functions may be delegated to hardware. The Link Controller is responsible for the processing of the baseband and the management of ARQ and physical layer FEC protocols. In addition, it handles the transfer functions (both asynchronous and synchronous), audio coding and data encryption. The CPU of the device is responsible for attending the instructions related to Bluetooth of the host device, in order to simplify its operation. To do this, the CPU runs software called Link Manager that has the function of communicating with other devices through the LMP protocol.

A Bluetooth device is a short-range wireless device. Bluetooth devices are fabricated on RF CMOS integrated circuit (RF circuit) chips.

## Bluetooth Protocol Stack

Bluetooth is defined as a layer protocol architecture consisting of core protocols, cable replacement protocols, telephony control protocols, and adopted protocols. Mandatory protocols for all Bluetooth stacks are LMP, L2CAP and SDP. In addition, devices that communicate with Bluetooth almost universally can use these protocols: HCI and RF-COMM.

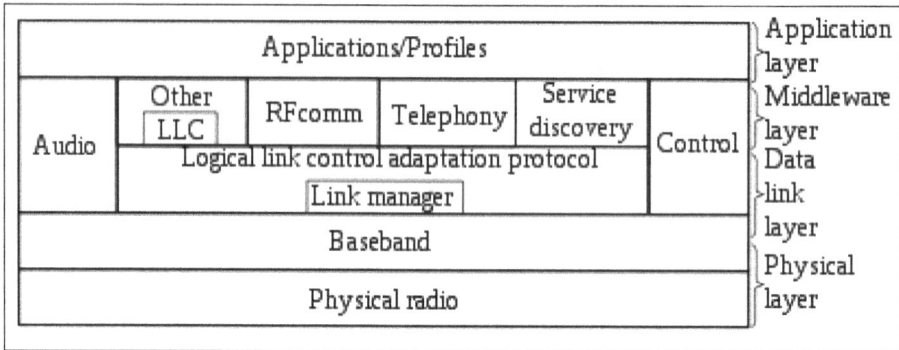

| | | | | | | |
|---|---|---|---|---|---|---|
| Applications/Profiles | | | | | | Application layer |
| Audio | Other LLC | RFcomm | Telephony | Service discovery | Control | Middleware layer |
| | Logical link control adaptation protocol | | | | | Data link layer |
| | Link manager | | | | | |
| Baseband | | | | | | |
| Physical radio | | | | | | Physical layer |

## Link Manager

The Link Manager (LM) is the system that manages establishing the connection between devices. It is responsible for the establishment, authentication and configuration of the link. The Link Manager locates other managers and communicates with them via the management protocol of the LMP link. In order to perform its function as a service provider, the LM uses the services included in the Link Controller (LC). The Link Manager Protocol basically consists of a number of PDUs (Protocol Data Units) that are sent from one device to another. The following is a list of supported services:

- Transmission and reception of data.

- Name request.

- Request of the link addresses.

- Establishment of the connection.

- Authentication.

- Negotiation of link mode and connection establishment.

## Host Controller Interface

The Host Controller Interface provides a command interface for the controller and for the link manager, which allows access to the hardware status and control registers. This interface provides an access layer for all Bluetooth devices. The HCI layer of the machine exchanges commands and data with the HCI firmware present in the Bluetooth device. One of the most important HCI tasks that must be performed is the automatic discovery of other Bluetooth devices that are within the coverage radius.

## Logical Link Control and Adaptation Protocol

The Logical Link Control and Adaptation Protocol (L2CAP) is used to multiplex multiple logical connections between two devices using different higher level protocols. Provides segmentation and reassembly of on-air packets.

In Basic mode, L2CAP provides packets with a payload configurable up to 64 kB, with 672 bytes as the default MTU, and 48 bytes as the minimum mandatory supported MTU.

In Retransmission and Flow Control modes, L2CAP can be configured either for isochronous data or reliable data per channel by performing retransmissions and CRC checks.

Bluetooth Core Specification Addendum 1 adds two additional L2CAP modes to the core specification. These modes effectively deprecate original Retransmission and Flow Control modes:

### Enhanced Retransmission Mode (ERTM)

This mode is an improved version of the original retransmission mode. This mode provides a reliable L2CAP channel.

### Streaming Mode (SM)

This is a very simple mode, with no retransmission or flow control. This mode provides an unreliable L2CAP channel.

Reliability in any of these modes is optionally and/or additionally guaranteed by the lower layer Bluetooth BDR/EDR air interface by configuring the number of retransmissions and flush timeout (time after which the radio flushes packets). In-order sequencing is guaranteed by the lower layer.

Only L2CAP channels configured in ERTM or SM may be operated over AMP logical links.

### Service Discovery Protocol

The Service Discovery Protocol (SDP) allows a device to discover services offered by other devices, and their associated parameters. For example, when you use a mobile phone with a Bluetooth headset, the phone uses SDP to determine which Bluetooth profiles the headset can use (Headset Profile, Hands Free Profile, Advanced Audio Distribution Profile (A2DP) etc.) and the protocol multiplexer settings needed for the phone to connect to the headset using each of them. Each service is identified by a Universally

Unique Identifier (UUID), with official services (Bluetooth profiles) assigned a short form UUID (16 bits rather than the full 128).

## Radio Frequency Communications

Radio Frequency Communications (RFCOMM) is a cable replacement protocol used for generating a virtual serial data stream. RFCOMM provides for binary data transport and emulates EIA-232 (formerly RS-232) control signals over the Bluetooth baseband layer, i.e., it is a serial port emulation.

RFCOMM provides a simple, reliable, data stream to the user, similar to TCP. It is used directly by many telephony related profiles as a carrier for AT commands, as well as being a transport layer for OBEX over Bluetooth.

Many Bluetooth applications use RFCOMM because of its widespread support and publicly available API on most operating systems. Additionally, applications that used a serial port to communicate can be quickly ported to use RFCOMM.

## Bluetooth Network Encapsulation Protocol

The Bluetooth Network Encapsulation Protocol (BNEP) is used for transferring another protocol stack's data via an L2CAP channel. Its main purpose is the transmission of IP packets in the Personal Area Networking Profile. BNEP performs a similar function to SNAP in Wireless LAN.

## Audio/Video Control Transport Protocol

The Audio/Video Control Transport Protocol (AVCTP) is used by the remote control profile to transfer AV/C commands over an L2CAP channel. The music control buttons on a stereo headset use this protocol to control the music player.

## Audio/Video Distribution Transport Protocol

The Audio/Video Distribution Transport Protocol (AVDTP) is used by the advanced audio distribution (A2DP) profile to stream music to stereo headsets over an L2CAP channel intended for video distribution profile in the Bluetooth transmission.

## Telephony Control Protocol

The Telephony Control Protocol – Binary (TCS BIN) is the bit-oriented protocol that defines the call control signaling for the establishment of voice and data calls between Bluetooth devices. Additionally, "TCS BIN defines mobility management procedures for handling groups of Bluetooth TCS devices".

TCS-BIN is only used by the cordless telephony profile, which failed to attract implementers. As such it is only of historical interest.

## Adopted Protocols

Adopted protocols are defined by other standards-making organizations and incorporated into Bluetooth's protocol stack, allowing Bluetooth to code protocols only when necessary. The adopted protocols include:

### Point-to-Point Protocol

Internet standard protocol for transporting IP datagrams over a point-to-point link.

### TCP/IP/UDP

Foundation Protocols for TCP/IP protocol suite.

### Object Exchange Protocol

Session-layer protocol for the exchange of objects, providing a model for object and operation representation.

### Wireless Application Environment/Wireless Application Protocol

WAE specifies an application framework for wireless devices and WAP is an open standard to provide mobile users access to telephony and information services.

### Baseband Error Correction

Depending on packet type, individual packets may be protected by error correction, either 1/3 rate forward error correction (FEC) or 2/3 rate. In addition, packets with CRC will be retransmitted until acknowledged by automatic repeat request (ARQ).

### Setting up Connections

Any Bluetooth device in discoverable mode transmits the following information on demand:

- Device name.
- Device class.
- List of services.
- Technical information (for example: device features, manufacturer, Bluetooth specification used, clock offset).

Any device may perform an inquiry to find other devices to connect to, and any device can be configured to respond to such inquiries. However, if the device trying to connect knows the address of the device, it always responds to direct connection requests and transmits the information shown in the list above if requested. Use of a device's services may require pairing or acceptance by its owner, but the connection itself can be initiated by any device and held until it goes out of range. Some devices can be connected to only one device at a time, and connecting to them prevents them from connecting to other devices and appearing in inquiries until they disconnect from the other device.

Every device has a unique 48-bit address. However, these addresses are generally not shown in inquiries. Instead, friendly Bluetooth names are used, which can be set by the user. This name appears when another user scans for devices and in lists of paired devices.

Most cellular phones have the Bluetooth name set to the manufacturer and model of the phone by default. Most cellular phones and laptops show only the Bluetooth names and special programs are required to get additional information about remote devices. This can be confusing as, for example, there could be several cellular phones in range named T610.

## Pairing and Bonding

### Motivation

Many services offered over Bluetooth can expose private data or let a connecting party control the Bluetooth device. Security reasons make it necessary to recognize specific devices, and thus enable control over which devices can connect to a given Bluetooth device. At the same time, it is useful for Bluetooth devices to be able to establish a connection without user intervention (for example, as soon as in range).

To resolve this conflict, Bluetooth uses a process called bonding, and a bond is generated through a process called pairing. The pairing process is triggered either by a specific request from a user to generate a bond (for example, the user explicitly requests to "Add a Bluetooth device"), or it is triggered automatically when connecting to a service where (for the first time) the identity of a device is required for security purposes. These two cases are referred to as dedicated bonding and general bonding respectively.

Pairing often involves some level of user interaction. This user interaction confirms the identity of the devices. When pairing successfully completes, a bond forms between the two devices, enabling those two devices to connect to each other in the future without repeating the pairing process to confirm device identities. When desired, the user can remove the bonding relationship.

### Implementation

During pairing, the two devices establish a relationship by creating a shared secret known as a link key. If both devices store the same link key, they are said to be paired or bonded. A device that wants to communicate only with a bonded device can cryptographically authenticate the identity of the other device, ensuring it is the same device it previously paired with. Once a link key is generated, an authenticated Asynchronous Connection-Less (ACL) link between the devices may be encrypted to protect exchanged data against eavesdropping. Users can delete link keys from either device, which removes the bond between the devices so it is possible for one device to have a stored link key for a device it is no longer paired with.

Bluetooth services generally require either encryption or authentication and as such require pairing before they let a remote device connect. Some services, such as the Object Push Profile, elect not to explicitly require authentication or encryption so that pairing does not interfere with the user experience associated with the service use-cases.

## Pairing Mechanisms

Pairing mechanisms changed significantly with the introduction of Secure Simple Pairing in Bluetooth v2.1. The following summarizes the pairing mechanisms:

- Legacy pairing: This is the only method available in Bluetooth v2.0 and before. Each device must enter a PIN code; pairing is only successful if both devices enter the same PIN code. Any 16-byte UTF-8 string may be used as a PIN code; however, not all devices may be capable of entering all possible PIN codes.

  ○ Limited input devices: The obvious example of this class of device is a Bluetooth Hands-free headset, which generally have few inputs. These devices usually have a fixed PIN, for example "0000" or "1234", that are hard-coded into the device.

  ○ Numeric input devices: Mobile phones are classic examples of these devices. They allow a user to enter a numeric value up to 16 digits in length.

  ○ Alpha-numeric input devices: PCs and smartphones are examples of these devices. They allow a user to enter full UTF-8 text as a PIN code. If pairing with a less capable device the user must be aware of the input limitations on the other device; there is no mechanism available for a capable device to determine how it should limit the available input a user may use.

- Secure Simple Pairing (SSP): This is required by Bluetooth v2.1, although a Bluetooth v2.1 device may only use legacy pairing to interoperate with a v2.0 or earlier device. Secure Simple Pairing uses a form of public key cryptography, and some types can help protect against man in the middle, or MITM attacks. SSP has the following authentication mechanisms:

  ○ Just works: As the name implies, this method just works, with no user interaction. However, a device may prompt the user to confirm the pairing process. This method is typically used by headsets with very limited IO capabilities, and is more secure than the fixed PIN mechanism this limited set of devices uses for legacy pairing. This method provides no man-in-the-middle (MITM) protection.

- ○ Numeric comparison: If both devices have a display, and at least one can accept a binary yes/no user input, they may use Numeric Comparison. This method displays a 6-digit numeric code on each device. The user should compare the numbers to ensure they are identical. If the comparison succeeds, the user(s) should confirm pairing on the device(s) that can accept an input. This method provides MITM protection, assuming the user confirms on both devices and actually performs the comparison properly.

- ○ Passkey Entry: This method may be used between a device with a display and a device with numeric keypad entry (such as a keyboard), or two devices with numeric keypad entry. In the first case, the display presents a 6-digit numeric code to the user, who then enters the code on the keypad. In the second case, the user of each device enters the same 6-digit number. Both of these cases provide MITM protection.

- ○ Out of band (OOB): This method uses an external means of communication, such as near-field communication (NFC) to exchange some information used in the pairing process. Pairing is completed using the Bluetooth radio, but requires information from the OOB mechanism. This provides only the level of MITM protection that is present in the OOB mechanism.

SSP is considered simple for the following reasons:

- In most cases, it does not require a user to generate a passkey.

- For use cases not requiring MITM protection, user interaction can be eliminated.

- For numeric comparison, MITM protection can be achieved with a simple equality comparison by the user.

- Using OOB with NFC enables pairing when devices simply get close, rather than requiring a lengthy discovery process.

## Security Concerns

Prior to Bluetooth v2.1, encryption is not required and can be turned off at any time. Moreover, the encryption key is only good for approximately 23.5 hours; using a single encryption key longer than this time allows simple XOR attacks to retrieve the encryption key.

- Turning off encryption is required for several normal operations, so it is problematic to detect if encryption is disabled for a valid reason or for a security attack.

Bluetooth v2.1 addresses this in the following ways:

- Encryption is required for all non-SDP (Service Discovery Protocol) connections.

- A new Encryption Pause and Resume feature is used for all normal operations that require that encryption be disabled. This enables easy identification of normal operation from security attacks.

- The encryption key must be refreshed before it expires.

Link keys may be stored on the device file system, not on the Bluetooth chip itself. Many Bluetooth chip manufacturers let link keys be stored on the device however, if the device is removable, this means that the link key moves with the device.

## HiperLAN

HiperLAN (High Performance Radio LAN) is a wireless LAN standard. It is a European alternative for the IEEE 802.11 standards (the IEEE is an international organization). It is defined by the European Telecommunications Standards Institute (ETSI). In ETSI the standards are defined by the BRAN project (Broadband Radio Access Networks). The HiperLAN standard family has four different versions.

## HiperLAN/1

Planning for the first version of the standard, called HiperLAN/1, started 1992, when planning of 802.11 was already going on. The goal of the HiperLAN was the high data rate, higher than 802.11. The standard was approved in 1997. The functional specification is EN300652, the rest is in ETS300836.

The standard covers the Physical layer and the Media Access Control part of the Data link layer like 802.11. There is a new sublayer called Channel Access and Control sublayer (CAC). This sublayer deals with the access requests to the channels. The accomplishing of the request is dependent on the usage of the channel and the priority of the request.

CAC layer provides hierarchical independence with Elimination-Yield Non-Preemptive Multiple Access mechanism (EY-NPMA). EY-NPMA codes priority choices and other functions into one variable length radio pulse preceding the packet data. EY-NPMA enables the network to function with few collisions even though there would be a large number of users. Multimedia applications work in HiperLAN because of EY-NPMA priority mechanism. MAC layer defines protocols for routing, security and power saving and provides naturally data transfer to the upper layers.

On the physical layer FSK and GMSK modulations are used in HiperLAN/1.

HiperLAN features:

- Ange 100 m.

- Slow mobility (1.4 m/s).

- Supports asynchronous and synchronous traffic.

- Bit rate 23.59 mbit/s.

- Description-wireless Ethernet.

- Frequency range- 5 ghz.

HiperLAN does not conflict with microwave and other kitchen appliances, which are on 2.4 GHz. An innovative feature of HIPERLAN 1, which other wireless networks do not offer, is its ability to forward data packets using several relays. Relays can extend the communication on the MAC layer beyond the radio range. For power conservation, a node may set up a specific wake up pattern. This pattern determines at what time the node is ready to receive, so that at other times, the node can turn off its receiver and save energy. These nodes are called p-savers and need so called p-supporters that contain information about wake up patterns of all the p-savers they are responsible for. A p-supporter only forwards data to a p-saver at the moment p-saver is awake. This action also requires buffering mechanisms for packets on p-supporting forwarders.

## HiperLAN/2

HiperLAN/2 functional specification was accomplished February 2000. Version 2 is designed as a fast wireless connection for many kinds of networks. Those are UMTS back bone network, ATM and IP networks. Also it works as a network at home like HiperLAN/1. HiperLAN/2 uses the 5 GHz band and up to 54 Mbit/s data rate.

The physical layer of HiperLAN/2 is very similar to IEEE 802.11a wireless local area networks. However, the media access control (the multiple access protocol) is Dynamic TDMA in HiperLAN/2, while CSMA/CA is used in 802.11a/n.

Basic services in HiperLAN/2 are data, sound, and video transmission. The emphasis is in the quality of these services (QoS).

The standard covers Physical, Data Link Control and Convergence layers. Convergence layer takes care of service dependent functionality between DLC and Network layer (OSI 3). Convergence sublayers can be used also on the physical layer to connect IP, ATM or UMTS networks. This feature makes HiperLAN/2 suitable for the wireless connection of various networks.

On the physical layer BPSK, QPSK, 16QAM or 64QAM modulations are used.

HiperLAN/2 offers security measures. The data are secured with DES or Triple DES algorithms. The wireless access point and the wireless terminal can authenticate each other.

Most important worldwide manufacturers of HiperLAN/2 are Alvarion (Israel), Freescale (USA), Panasonic (Japan).

## Wireless PAN

A personal area network (PAN) is a computer network for interconnecting devices centered on an individual person's workspace.A PAN provides data transmission among devices such as computers, smartphones, tablets and personal digital assistants. PANs can be used for communication among the personal devices themselves, or for connecting to a higher level network and the Internet where one master device takes up the role as gateway. A PAN may be wireless or carried over wired interfaces such as USB.

A wireless personal area network (WPAN) is a PAN carried over a low-powered, short-distance wireless network technology such as IrDA, Wireless USB, Bluetooth or ZigBee. The reach of a WPAN varies from a few centimeters to a few meters.

## Wired Personal Area Network

Wired personal area networks provide short connections between peripherals. Example technologies include:

- USB.

- IEEE-1394.

- Thunderbolt (interface).

## Wireless Personal Area Network

A wireless personal area network (WPAN) is a personal area network in which the connections are wireless. IEEE 802.15 has produced standards for several types of PANs operating in the ISM band including Bluetooth. The Infrared Data Association has produced standards for WPANs which operate using infrared communications.

## Bluetooth

Bluetooth uses short-range radio waves. Uses in a WPAN include, for example, Bluetooth devices such as keyboards, pointing devices, audio headsets, and printers may connect to smartwatches, cell phones, or computers. A Bluetooth WPAN is also called a piconet, and is composed of up to 8 active devices in a master-slave relationship (a very large number of additional devices can be connected in "parked" mode). The first

Bluetooth device in the piconet is the master, and all other devices are slaves that communicate with the master. A piconet typically has a range of 10 metres (33 ft.), although ranges of up to 100 metres (330 ft.) can be reached under ideal circumstances. Long-range Bluetooth routers with augmented antenna arrays connect Bluetooth devices up to 1,000 feet.

With Bluetooth mesh networking the range and number of devices is extended by using mesh networking techniques to relay information from one to another. Such a network doesn't have a master device and may or may not be treated as a WPAN.

## IrDA

Infrared Data Association (IrDA) uses infrared light, which has a frequency below the human eye's sensitivity. Infrared is used in other wireless communications applications, for instance, in remote controls. Typical WPAN devices that use IrDA include printers, keyboards, and other serial communication interfaces.

## Wireless LAN

A wireless LAN (WLAN) is a wireless computer network that links two or more devices using wireless communication to form a local area network (LAN) within a limited area such as a home, school, computer laboratory, campus, or office building. This gives users the ability to move around within the area and remain connected to the network. Through a gateway, a WLAN can also provide a connection to the wider Internet.

Most modern WLANs are based on IEEE 802.11 standards and are marketed under the Wi-Fi brand name.

Wireless LANs have become popular for use in the home, due to their ease of installation and use. They are also popular in commercial properties that offer wireless access to their employees and customers.

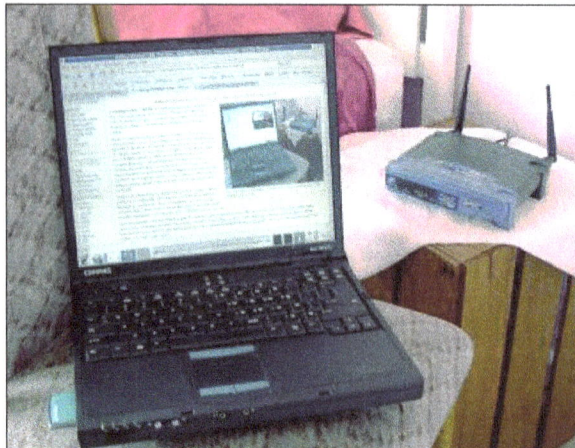

This notebook computer is connected to a wireless access point using a PC Card wireless card.

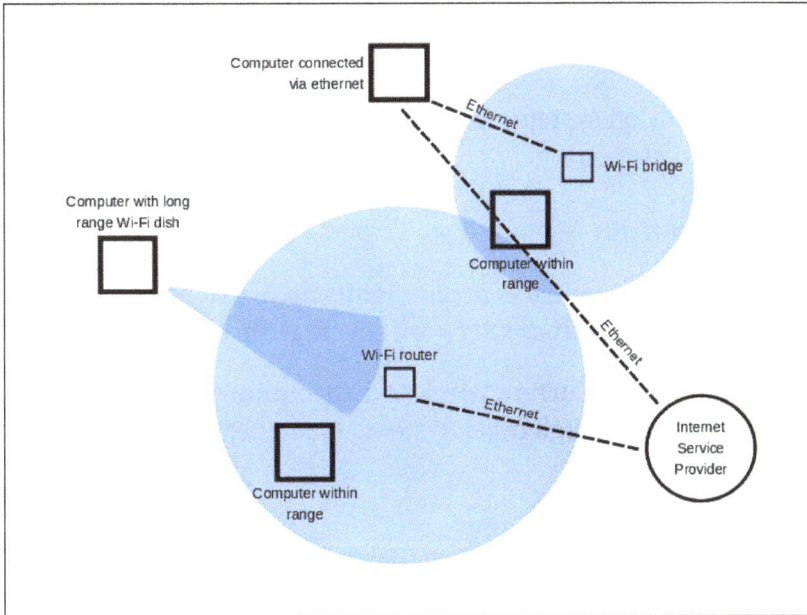

An example of a Wi-Fi network.

## Architecture

### Stations

All components that can connect into a wireless medium in a network are referred to as stations (STA). All stations are equipped with wireless network interface controllers (WNICs). Wireless stations fall into two categories: wireless access points, and clients. Access points (APs), normally wireless routers, are base stations for the wireless network. They transmit and receive radio frequencies for wireless enabled devices to communicate with. Wireless clients can be mobile devices such as laptops, personal digital assistants, IP phones and other smartphones, or non-portable devices such as desktop computers, printers, and workstations that are equipped with a wireless network interface.

### Basic Service Set

The basic service set (BSS) is a set of all stations that can communicate with each other at PHY layer. Every BSS has an identification (ID) called the BSSID, which is the MAC address of the access point servicing the BSS.

There are two types of BSS: Independent BSS (also referred to as IBSS), and infrastructure BSS. An independent BSS (IBSS) is an ad hoc network that contains no access points, which means they cannot connect to any other basic service set.

### Independent Basic Service Set

An IBSS is a set of STAs configured in ad hoc (peer-to-peer)mode.

## Extended Service Set

An extended service set (ESS) is a set of connected BSSs. Access points in an ESS are connected by a distribution system. Each ESS has an ID called the SSID which is a 32-byte (maximum) character string.

## Distribution System

A distribution system (DS) connects access points in an extended service set. The concept of a DS can be used to increase network coverage through roaming between cells.

DS can be wired or wireless. Current wireless distribution systems are mostly based on WDS or MESH protocols, though other systems are in use.

## Types of Wireless LANs

The IEEE 802.11 has two basic modes of operation: infrastructure and ad hoc mode. In ad hoc mode, mobile units transmit directly peer-to-peer. In infrastructure mode, mobile units communicate through an access point that serves as a bridge to other networks (such as Internet or LAN).

Since wireless communication uses a more open medium for communication in comparison to wired LANs, the 802.11 designers also included encryption mechanisms: Wired Equivalent Privacy (WEP, now insecure), Wi-Fi Protected Access (WPA, WPA2, WPA3), to secure wireless computer networks. Many access points will also offer Wi-Fi Protected Setup, a quick (but now insecure) method of joining a new device to an encrypted network.

## Infrastructure

Most Wi-Fi networks are deployed in infrastructure mode.

In infrastructure mode, a base station acts as a wireless access point hub, and nodes communicate through the hub. The hub usually, but not always, has a wired or fiber network connection, and may have permanent wireless connections to other nodes.

Wireless access points are usually fixed, and provide service to their client nodes within range.

Wireless clients, such as laptops and smartphones, connect to the access point to join the network.

Sometimes a network will have a multiple access points, with the same 'SSID' and security arrangement. In that case connecting to any access point on that network joins the client to the network. In that case, the client software will try to choose the access point to try to give the best service, such as the access point with the strongest signal.

## Peer-to-peer

An ad hoc network (not the same as a WiFi Direct network) is a network where stations communicate only peer to peer (P2P). There is no base and no one gives permission to talk. This is accomplished using the Independent Basic Service Set (IBSS).

A WiFi Direct network is another type of network where stations communicate peer to peer.

In a Wi-Fi P2P group, the group owner operates as an access point and all other devices are clients. There are two main methods to establish a group owner in the Wi-Fi Direct group. In one approach, the user sets up a P2P group owner manually. This method is also known as Autonomous Group Owner (autonomous GO). In the second method, also called negotiation-based group creation, two devices compete based on the group owner intent value. The device with higher intent value becomes a group owner and the second device becomes a client. Group owner intent value can depend on whether the wireless device performs a cross-connection between an infrastructure WLAN service and a P2P group, remaining power in the wireless device, whether the wireless device is already a group owner in another group and/or a received signal strength of the first wireless device.

A peer-to-peer network allows wireless devices to directly communicate with each other. Wireless devices within range of each other can discover and communicate directly without involving central access points. This method is typically used by two computers so that they can connect to each other to form a network. This can basically occur in devices within a closed range.

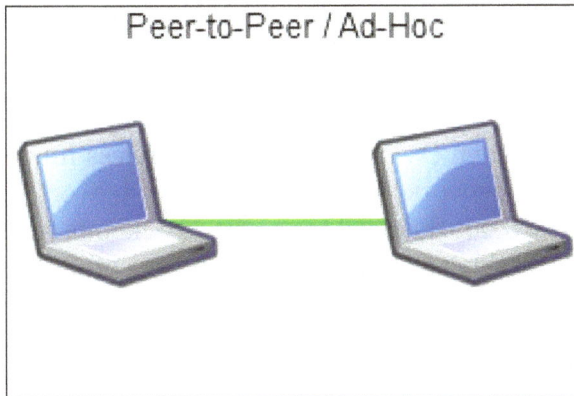

Peer-to-Peer / Ad-Hoc

Peer-to-Peer or ad hoc wireless LAN.

If a signal strength meter is used in this situation, it may not read the strength accurately and can be misleading, because it registers the strength of the strongest signal, which may be the closest computer.

IEEE 802.11 defines the physical layer (PHY) and MAC (Media Access Control) layers based on CSMA/CA (Carrier Sense Multiple Access with Collision Avoidance). This is in contrast to Ethernet which uses CSMA-CD (Carrier Sense Multiple Access with Collision Detection). The 802.11 specification includes provisions designed to minimize collisions, because two mobile units may both be in range of a common access point, but out of range of each other.

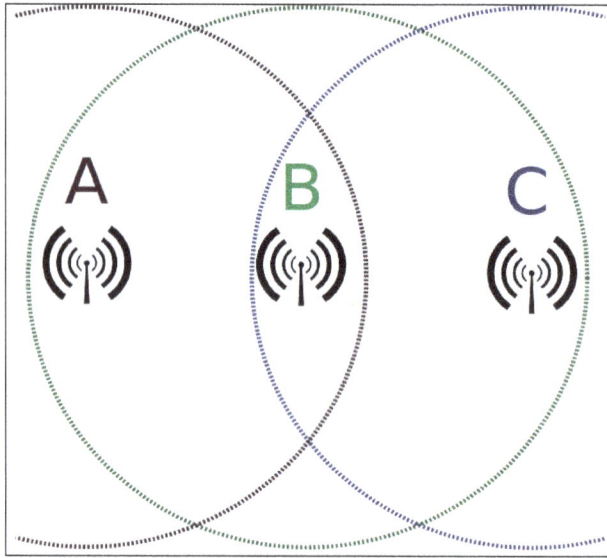

Hidden node problem: Devices A and C are both communicating with B, but are unaware of each other.

## Bridge

A bridge can be used to connect networks, typically of different types. A wireless Ethernet bridge allows the connection of devices on a wired Ethernet network to a wireless network. The bridge acts as the connection point to the Wireless LAN.

## Wireless Distribution System

A wireless distribution system (WDS) enables the wireless interconnection of access points in an IEEE 802.11 network. It allows a wireless network to be expanded using multiple access points without the need for a wired backbone to link them, as is traditionally required. The notable advantage of a WDS over other solutions is that it preserves the MAC addresses of client packets across links between access points.

An access point can be either a main, relay or remote base station. A main base station is typically connected to the wired Ethernet. A relay base station relays data between remote base stations, wireless clients or other relay stations to either a main or another relay base station. A remote base station accepts connections from wireless clients and passes them to relay or main stations. Connections between clients are made using MAC addresses rather than by specifying IP assignments.

All base stations in a WDS must be configured to use the same radio channel, and share WEP keys or WPA keys if they are used. They can be configured to different service set identifiers. WDS also requires that every base station be configured to forward to others in the system as mentioned above.

WDS capability may also be referred to as repeater mode because it appears to bridge and accept wireless clients at the same time (unlike traditional bridging). Throughput in this method is halved for all clients connected wirelessly.

When it is difficult to connect all of the access points in a network by wires, it is also possible to put up access points as repeaters.

## Roaming

There are two definitions for wireless LAN roaming:

- Internal roaming: The Mobile Station (MS) moves from one access point (AP) to another AP within a home network if the signal strength is too weak. An authentication server (RADIUS) performs the re-authentication of MS via 802.1x (e.g. with PEAP). The billing of QoS is in the home network. A Mobile Station roaming from one access point to another often interrupts the flow of data among the Mobile Station and an application connected to the network. The Mobile Station, for instance, periodically monitors the presence of alternative access points (ones that will provide a better connection). At some point, based on proprietary mechanisms, the Mobile Station decides to re-associate with an access point having a stronger wireless signal. The Mobile Station, however, may lose a connection with an access point before associating with another access point. In order to provide reliable connections with applications, the Mobile Station must generally include software that provides session persistence.

- External roaming: The MS (client) moves into a WLAN of another Wireless Internet Service Provider (WISP) and takes their services (Hotspot). The user can use a foreign network independently from their home network, provided that the foreign network allows visiting users on their network. There must be special authentication and billing systems for mobile services in a foreign network.

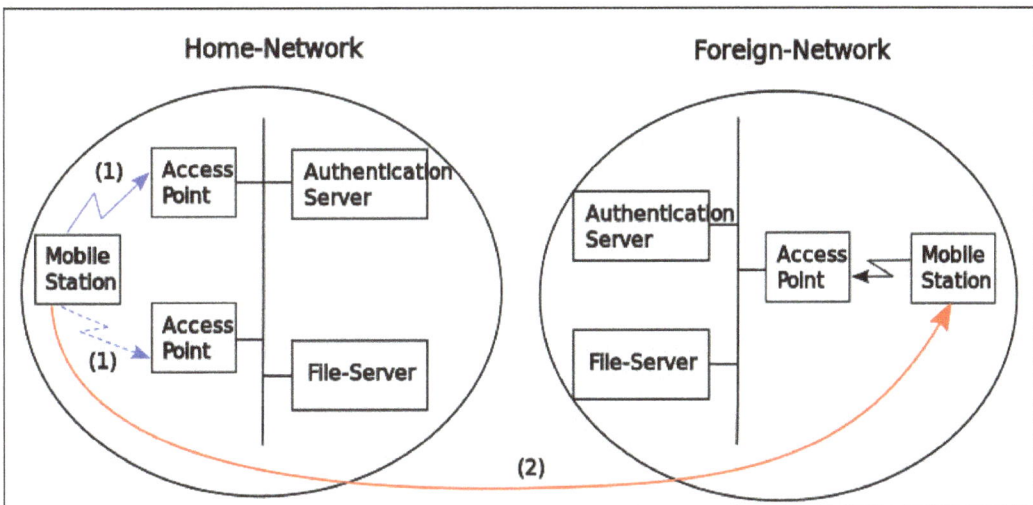

## Applications

Wireless LANs have a great deal of applications. Modern implementations of WLANs range from small in-home networks to large, campus-sized ones to completely mobile networks on airplanes and trains.

Users can access the Internet from WLAN hotspots in restaurants, hotels, and now with portable devices that connect to 3G or 4G networks. Oftentimes these types of public access points require no registration or password to join the network. Others can be accessed once registration has occurred and/or a fee is paid.

Existing Wireless LAN infrastructures can also be used to work as indoor positioning systems with no modification to the existing hardware.

## Wireless MAN

A metropolitan area network (MAN) is a network that interconnects users with computer resources in a geographic area or region larger than that covered by even a large local area network (LAN) but smaller than the area covered by a wide area network (WAN). The term is applied to the interconnection of networks in a city into a single larger network (which may then also offer efficient connection to a wide area network). It is also used to mean the interconnection of several local area networks by bridging them with backbone lines. The latter usage is also sometimes referred to as a campus network.

Examples of metropolitan area networks of various sizes can be found in the metropolitan areas of London, England; Lodz, Poland; and Geneva, Switzerland. Large universities also sometimes use the term to describe their networks. A recent trend is the installation of wireless MANs.

## Wireless WAN

Wireless wide area network (WWAN), is a form of wireless network. The larger size of a wide area network compared to a local area network requires differences in technology. Wireless networks of different sizes deliver data in the form of telephone calls, web pages, and streaming video.

A WWAN often differs from wireless local area network (WLAN) by using mobile telecommunication cellular network technologies such as 2G, 3G, 4G LTE, and 5G to transfer data. It is sometimes referred as Mobile Broadband. These technologies are offered regionally, nationwide, or even globally and are provided by a wireless service provider. WWAN connectivity allows a user with a laptop and a WWAN card to surf the web, check email, or connect to a virtual private network (VPN) from anywhere within the regional boundaries of cellular service. Various computers can have integrated WWAN capabilities.

A WWAN may also be a closed network that covers a large geographic area. For example, a mesh network or MANET with nodes on buildings, towers, trucks, and planes could also be considered a WWAN.

A WWAN may also be a low-power, low-bit-rate wireless WAN, (LPWAN), intended to carry small packets of information between things, often in the form of battery operated sensors.

Since radio communications systems do not provide a physically secure connection path, WWANs typically incorporate encryption and authentication methods to make them more secure. Some of the early GSM encryption techniques were flawed, and security experts have issued warnings that cellular communication, including WWAN, is no longer secure. UMTS (3G) encryption was developed later and has yet to be broken.

# Wireless Application Protocol

The Wireless Application Protocol (WAP) is a worldwide standard for the delivery and presentation of wireless information to mobile phones and other wireless devices. The idea behind WAP is simple: simplify the delivery of Internet content to wireless devices by delivering a comprehensive, Internet-based, wireless specification. The WAP Forum released the first version of WAP in 1998. Since then, it has been widely adopted by wireless phone manufacturers, wireless carriers, and application developers worldwide. Many industry analysts estimate that 90 percent of mobile phones sold over the next few years will be WAP-enabled.

The WAP architecture is composed of various protocols and an XML-based markup language called the Wireless Markup Language (WML), which is the successor to the Handheld Device Markup Language (HDML) as defined by Openwave Systems. WAP 2.x contains a new version of WML, commonly referred to as WML2; it is based on the eXtensible HyperText Markup Language (XHTML), signaling part of WAP's move toward using common Internet specifications such as HTTP and TCP/IP.

## WAP Programming Model

The WAP programming model is very similar to the Internet programming model. It typically uses the pull approach for requesting content, meaning the client makes the request for content from the server. However, WAP also supports the ability to push content from the server to the client using the Wireless Telephony Application Specification (WTA), which provides the ability to access telephony functions on the client device.

Content can be delivered to a wireless device using WAP in two ways: with or without a WAP gateway. Whether a gateway is used depends on the features required and the version of WAP being implemented. WAP 1.x requires the use of a WAP gateway as an intermediary between the client and the wireless application server. This gateway is responsible for the following:

- Translating requests from the WAP protocol to the protocols used over the World Wide Web, such as HTTP and TCP/IP.

- Encoding and decoding regular Web content into compact formats that are more appropriate for wireless communication.

- Allowing use of standard HTTP-based Web servers for the generation and delivery of wireless content. This may involve transforming the content to make it appropriate for wireless consumption.

- Implementing push functionality using WTA.

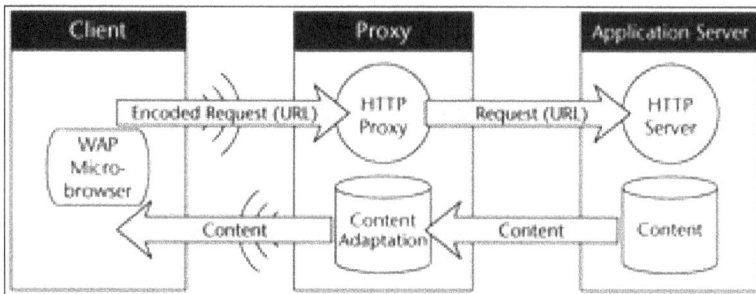

WAP Programming model using a wireless gateway (or proxy).

The WAP gateway is often called the WAP proxy in the WAP 2.x documents available from the OMA.

When developing WAP 2.x applications, you no longer are required to use a WAP gateway. WAP 2.x allows HTTP communication between the client and the origin server, so there is no need for conversion. This is not to say, however, that a WAP gateway is not beneficial. Using a WAP gateway will allow you to optimize the communication process and facilitate other wireless service features such as location, privacy, and WAP Push. The WAP programming model without a WAP gateway: Note that removing it makes the wireless Internet application architecture nearly identical to that used for standard Web applications.

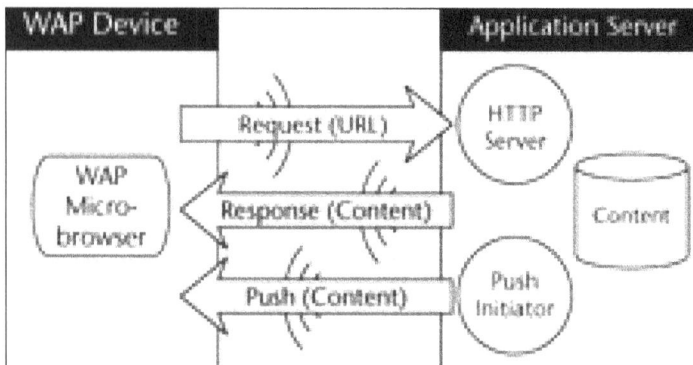

WAP programming model without gateway.

Both WAP programming models require the same core set of steps to process a wireless Internet request. These steps are based on the common pull model used for Internet applications; that is, a request/response method for communication.

## WAP Components

The WAP architecture comprises several components, each serving a specific function. These components include a wireless application environment, session and transaction support, security, and data transfer. The exact protocols used depend on which version of WAP you are implementing. WAP 2.x is based mainly on common Internet protocols such as HTTP and TCP/IP, while WAP 1.x uses proprietary protocols developed as part of the WAP specification. We will investigate each component and its related function.

To begin, we will look at how WAP conforms to the Open Systems Interconnection (OSI) model as defined by the International Standards Organization (ISO). The OSI model consists of seven distinct layers, six of which are depicted in figure as they relate to the WAP architecture. The physical layer is not shown; it sits below the network layer and defines the physical aspects such as the hardware and the raw bit-stream. For each of the other six layers, WAP has a corresponding layer.

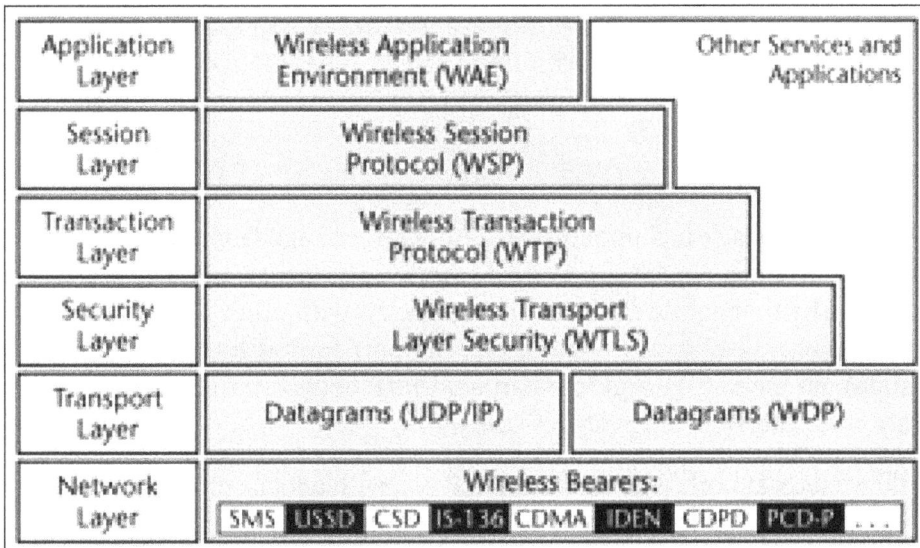

| Application Layer | Wireless Application Environment (WAE) | Other Services and Applications |
| --- | --- | --- |
| Session Layer | Wireless Session Protocol (WSP) | |
| Transaction Layer | Wireless Transaction Protocol (WTP) | |
| Security Layer | Wireless Transport Layer Security (WTLS) | |
| Transport Layer | Datagrams (UDP/IP) | Datagrams (WDP) |
| Network Layer | Wireless Bearers: SMS USSD CSD IS-136 CDMA IDEN CDPD PCD-P ... | |

WAP architecture and its relationship to the OSI model.

## Wireless Application Environment

The Wireless Application Environment (WAE) is the application layer of the OSI model. It provides the required elements for interaction between Web applications and wireless clients using a WAP microbrowser. These elements are as follows:

- A specification for a microbrowser that controls the user interface and interprets WML and WMLScript.

- The foundation for the microbrowser in the form of the Wireless Markup Language (WML). WML has been designed to accommodate the unique characteristics of wireless devices, by incorporating a user interface model that is suitable for small form-factor devices that do not have a QWERTY keyboard.

- A complete scripting language called WMLScript that extends the functionality of WML, enabling more capabilities on the client for business and presentation logic.

- Support for other content types such as wireless bitmap images (WBMP), vCard, and vCalendar.

WAP 2.x extends WAE by adding the following elements:

- A new markup language specification called WML2 that is based on XHT-ML-Basic. Backward compatibility with WML1 has been maintained.

- Support for stylesheets to enhance presentation capabilities. Stylesheet support is based on the Mobile Profile of Cascading Style Sheets (CSS) from the W3C, and supports both inline and external style sheets.

WAP 2.x WAE has backward compatibility to WML1. This is accomplished either via built-in support for both languages or by translating WML1 into WML2 using eXtensible Stylesheet Language Transformation (XSLT). The method used depends on the implementation by the device manufacturer.

## WAP Protocol Stack

The WAP protocol stack has undergone significant change from WAP 1.x to WAP 2.x. The basis for the change is the support for Internet Protocols (IPs) when IP connectivity is supported by the mobile device and network. As with other parts of WAP, the WAP 2.x protocol stack is backward-compatible. Support for the legacy WAP 1.x stack has been maintained for non-IP and low-bandwidth IP networks that can benefit from the optimizations in the WAP 1.x protocol stack.

We will take a look at both WAP 1.x and WAP 2.x, with a focus on the technologies used in each version of the specification.

## WAP 1.x

The protocols in the WAP 1.x protocol stack have been optimized for low-bandwidth, high-latency networks, which are prevalent in pre-3G wireless networks. The protocols are as follows:

Wireless Session Protocol (WSP). WSP provides capabilities similar to HTTP/1.1 while incorporating features designed for low-bandwidth, high-latency wireless networks such as long-lived sessions and session suspend/resume. This is particularly important, as it makes it possible to suspend a session while not in use, to free up network resources or preserve battery power. The communication from a WAP gateway to the microbrowser client is over WSP.

Wireless Transaction Protocol (WTP). WTP provides a reliable transport mechanism for the WAP datagram service. It offers similar reliability as Transmission Control Protocol/Internet Protocol (TCP/IP), but it removes characteristics that make TCP/IP unsuitable for wireless communication, such as the extra handshakes and additional information for handling out-of-order packets. Since the communication is directly from a handset to a server, this information is not required. The result is that WTP requires less than half of the number of packets of a standard HTTP-TCP/IP request. In addition, using WTP means that a TCP stack is not required on the wireless device, reducing the processing power and memory required.

Wireless Transport Layer Security (WTLS). WTLS is the wireless version of the Transport Security Layer (TLS), which was formerly known as Secure Sockets Layer (SSL). It provides privacy, data integrity, and authentication between the client and the wireless server. Using WTLS, WAP gateways can automatically provide wireless security for Web applications that use TLS. In addition, like the other wireless protocols, WTLS incorporates features designed for wireless networks, such as datagram support, optimized handshakes, and dynamic key refreshing.

Wireless Datagram Protocol (WDP). WDP is a datagram service that brings a common interface to wireless transportation bearers. It can provide this consistent layer by using a set of adapters designed for specific features of these bearers. It supports CDPD, GSM, CDMA, TDMA, SMS, FLEX (a wireless technology developed by Motorola), and Integrated Digital Enhanced Network (iDEN) protocols.

## WAP 2.x

One of the main new features in WAP 2.x is the use of Internet protocols in the WAP protocol stack. This change was precipitated by the rollout of 2.5G and 3G networks that provide IP support directly to wireless devices. To accommodate this change, WAP 2.x has the following new protocol layers.

Wireless Profiled HTTP (WP-HTTP). WP-HTTP is a profile of HTTP designed for the wireless environment. It is fully interoperable with HTTP/1.1 and allows the usage of the HTTP request/response model for interaction between the wireless device and the wireless server.

Transport Layer Security (TLS). WAP 2.0 includes a wireless profile of TLS, which allows secure transactions. The TLS profile includes cipher suites, certificate formats, signing algorithms, and the use of session resume, providing robust wireless security. There is also support for TLS tunneling, providing end-to-end security at the transport level. The support for TLS removes the WAP security gap that was present in WAP 1.x.

Wireless Profiled TCP (WP-TCP). WP-TCP is fully interoperable with standard Internet-based TCP implementations, while being optimized for wireless environments. These optimizations result in lower overhead for the communication stream.

Wireless devices can support both the WAP 1.x and WAP 2.x protocol stacks. In this scenario, they would need to operate independently of each other, since WAP 2.x provides support for both stacks.

## Other WAP 2.x Services

In addition to a new protocol stack, WAP 2.x introduced many other new features and services. These new features expand the capabilities of wireless devices and allow developers to create more useful applications and services. The following is a summary of the features of interest:

- WAP Push: WAP Push enables enterprises to initiate the sending of information on the server using a push proxy. This capability was introduced in WAP 1.2, but has been enhanced in WAP 2.x. Applications that require updates based on external information are particularly suited for using WAP Push. Examples include various forms of messaging applications, stock updates, airline departure and arrival updates, and traffic information. Before WAP Push was introduced, the wireless user was required to poll the server for updated information, wasting both time and bandwidth.

- User Agent Profile (UAProf): The UAProf enables a server to obtain information about the client making the request. In WAP 2.x, it is based on the Composite Capabilities/Preference Profiles (CC/PP) specification as defined by the W3C. It works by sending information in the request object, allowing wireless servers to adapt the information being sent according to the client device making the request.

- External Functionality Interface (EFI): This allows the WAP applications within the WAE to communicate with external applications, enabling other applications to extend the capabilities of WAP applications, similar to plug-ins for desktop browsers.

- Wireless Telephony Application (WTA): The WTA allows WAP applications to control various telephony applications, such as making calls, answering calls, putting calls on hold, or forwarding them. It allows WAP WTA-enabled cell phones to have integrated voice and data services.

- Persistent storage interface: WAP 2.x introduces a new storage service with a well-defined interface to store data locally on the device. The interface defines ways to organize, access, store, and retrieve data.

- Data synchronization: For data synchronization, WAP 2.x has adopted the SyncML solution.

- Multimedia Messaging Service (MMS): MMS is the framework for rich-content messaging. Going beyond what is possible for SMS, MMS can be used to transmit multimedia content such as pictures and videos. In addition, it can work with WAP Push and UAProf to send messages adapted specifically for the target client device.

## WAP Benefits

The WAP specification is continually changing to meet the growing demands of wireless applications. The majority of wireless carriers and handset manufacturers support WAP and continue to invest in the new capabilities it offers. Over the years WAP has evolved from using proprietary protocols in WAP 1.x to using standard Internet protocols in WAP 2.x, making it more approachable for Web developers. The following are some of the key benefits that WAP provides:

- WAP supports legacy WAP 1.x protocols that encode and optimize content for low-bandwidth, high-latency networks while communicating with the enterprise servers using HTTP.

- WAP supports wireless profiles of Internet protocols for interoperability with Internet applications. This allows WAP clients to communicate with enterprise servers, without requiring a WAP gateway.

- WAP allows end users to access a broad range of content over multiple wireless networks using a common user interface, the WAP browser. Because the WAP specification defines the markup language and microbrowser, users can be assured that wireless content will be suitable for their WAP-enabled device.

- WAP uses XML as the base language for both WML and WML2 (which uses XHTML), making it easy for application developers to learn and build wireless Internet applications. It also makes content transformation easier by incorporating support for XSL stylesheets to transform XML content. Once an application is developed using WML or WML2, any device that is WAP-compliant can access it.

- WAP has support for WTA. This allows applications to communicate with the device and network telephony functions. This permits the development of truly integrated voice and data applications.

- Using UAProf, the information delivered to each device can be highly customized.

- WAP works with all of the main wireless bearers, including CDPD, GSM, CDMA, TDMA, FLEX, and iDEN protocols. This interoperability allows developers to focus on creating their applications, without having to worry about the underlying network that will be used.

## WAP Gateway

(Wireless Application Protocol gateway) Software that decodes and encodes requests and responses between the smartphone micro browsers and the Internet. It decodes the encoded WAP requests from the micro browser and sends the HTTP requests to the Internet or to a local application server. It is the key component to) make mobile phones access the Internet. Since wireless environment brings a lot of limits on network features, the gateway is also used to enhance the performance of the network communication. The implementation of my WAP Gateway can integrate the security layer (WTLS) that provided by YanLiu's project to guarantee the security feature of the communication system between wireless clients and the WAP Gateway.

Mobile phone users have been quickly increasing these years and they have become a popular client system in the network. The techniques to get mobile phones access Internet are more and more popular. There are some features of wireless clients quite different from PC clients that require people to carefully to think about when implementing those techniques. Wireless clients usually have:

- Less powerful CPUs.

- Less memory.

- More latency.

- Less bandwidth.

WAP Gateway is a solution to these limits in wireless environment. It works with wireless clients as a Server in WSP (Wireless Session Protocols) domain and works with HTTP servers as a Client in HTTP domain. My implementation of WAP gateway addresses the last two limits – long latency and low bandwidth. It provides a feasible and efficient way to help wireless users communicate with HTTP servers.

### WAP Gateway Architecture

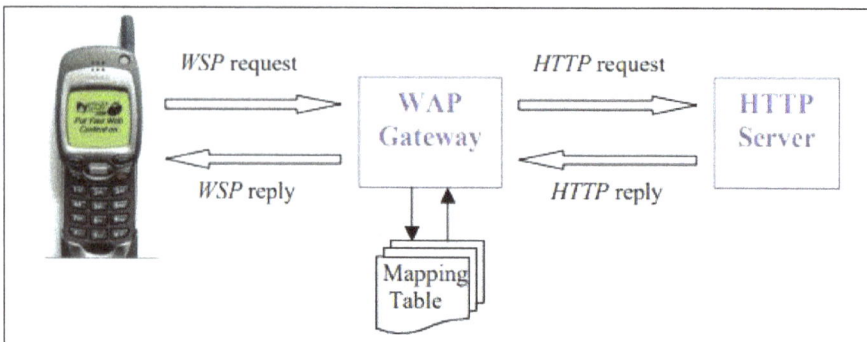

Wireless users usually request WML (Wireless Markup Language) data (usually a.wml file) from the HTTP server since wireless browsers are implemented to view WML format data. The data transfer procedure is as follows:

- Client sends a WSP request to WAP gateway.

- WAP gateway decodes the WSP request into HTTP request.

- WAP gateway sends the HTTP request to HTTP server.

- WAP gateway receives the HTTP reply from HTTP server.

- WAP gateway encodes the HTTP reply headers into WSP reply headers.

- WAP gateway uses WML compiler to encode the received WML data into WMLC format, which is a more compact format of data.

- WAP gateway sends WSP reply (WSP headers + WMLC data) to the client.

- Client parses WSP reply and presents the data.

## References

- Wireless-network: home-network-help.com, Retrieved 16 April, 2020

- Metropolitan-area-network-MAN: searchnetworking.techtarget.com

- Wolter Lemstra; Vic Hayes; John Groenewegen (December 27, 2010). The innovation journey of Wi-Fi: the road to global success. Cambridge University Press. p. 432. ISBN 0-521-19971-9

- Wireless-Application-Protocol-WAP-Overview,Mobile-devices-mobile-wireless-design-Part-Three-Building-Wireless-Internet-Applications- Thin-Client-Overview: etutorials.org, Retrieved 19 March, 2020

- Gratton, Dean A. (2013). The Handbook of Personal Area Networking Technologies and Protocols. Cambridge University Press. pp. 15–18. ISBN 9780521197267. Retrieved 12 December 2020

# Wireless Ad-Hoc Networks

# 4

- **Multicasting in Ad-Hoc Networks**
- **Mobile Ad-Hoc Networks**
- **Routing in Mobile Ad-Hoc Networks**

Wireless ad-hoc network is a decentralized wireless network that does not depend on routers and access points. Multicasting, routing and mobile ad-hoc networks are studied under its domain. This chapter has been carefully written to provide an easy under-standing of wireless ad hoc networks.

Wireless networks can be classified in two types: infrastructure network and in-frastructure less (ad-hoc) networks. Infrastructure network consists of a network with fixed and wired gateways. Ad-hoc means "for this or for this only". An ad hoc network is made up of multiple "nodes" connected by "links. Nodes can be the form of systems or devices i.e. mobile phone, laptop, personal digital assistance, MP3 player and personal computer that are participating in the networks. An ad hoc network typically refers to any set of networks where all devices have equal status on a network and are free to associate with any other ad hoc network device in link range. Ad-hoc network often refers to a mode of operation of IEEE 802.11 wireless networks. The three common sub types of these networks are Mobile Ad-hoc Net-works (MANETs), Wireless Sensor Networks and Wireless Mesh Networks. Ad-hoc networks have been proposed as an appealing communication technology to deal with the unexpected conditions emerged. Ad-hoc networks help to reduce adminis-trative cost. Wireless ad hoc network is a special structure of the wireless communi-cation network, whose communication relies on their cooperation among the nodes and achieves it in the manner of wireless multi-hop. Therefore, this kind of network does not rely on any fixed infrastructure, and has the properties of self-organizing and self-managing.

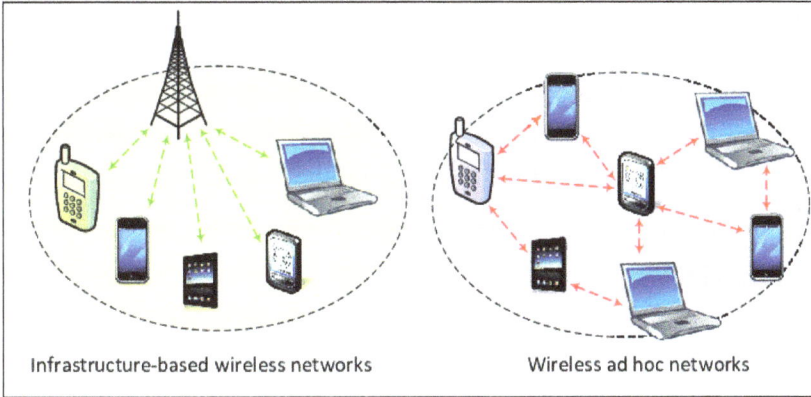

| Infrastructure-based wireless networks | Wireless ad hoc networks |

## Types of Wireless Ad-Hoc Networks

As we seen in introduction, classification of Wireless networks are infrastructure network and infrastructure less (ad hoc) networks and According to their application types of Wireless ad hoc networks are:

## Mobile Ad-Hoc Network

A mobile ad-hoc network is a self-configuring infrastructure less network of mobile devices connected by wireless links. Where no centralize authority is available and all nodes are independently follow the routine of mobility. Due to absence of the centralize authority the attacker easily join the network and perform the malicious activity. Beyond this gap to overcome this fault or problem various trust based security architecture are proposed for MANET. In general two types of communications can be considered in classical MANETs, broadcast communications and multihop communications via routing protocols.

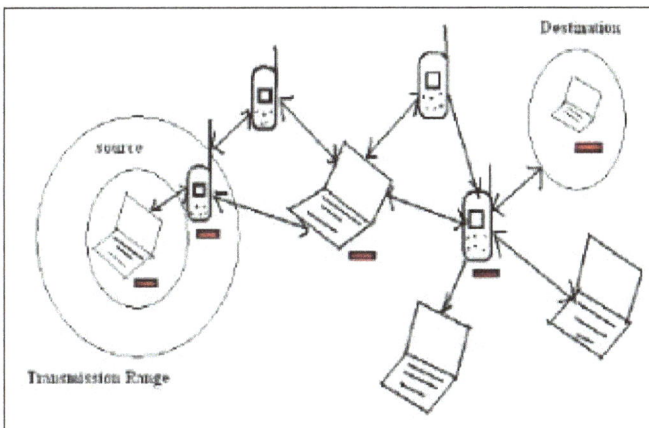

Mobile ad-hoc network.

The mobile ad hoc network has typical features namely Unreliability of wireless links between nodes and constantly changing topology. Each device in a MANET is free to move independently in any direction, and will therefore change its links to other devices frequently. Each must forward traffic unrelated to its own use, and therefore be a router. The primary challenge in building a MANET is equipping each device to continuously maintain the information required to properly route traffic. Such networks may operate by themselves or may be connected to the larger Internet. The achievement of MANET is hug growth of laptops and wireless or Wi- Fi networking.

## Types of MANET

Mobile Ad Hoc Networks are further classified into following three types:

### Vehicular Ad-Hoc Network

Vehicular Ad hoc Network (VANET), a subclass of mobile Ad Hoc networks (MANETs). Vehicular Ad Hoc Networks is a class of special wireless ad hoc network with the characteristics of high node mobility and fast topology changes. A Vehicular Ad-Hoc Network is a technology that uses moving cars as nodes in a network to create a mobile network. VANET turns every participating car into a wireless router or node, allowing cars approximately 100 to 300 metres of each other to connect and, in turn, create a network with a wide range. As cars fall out of the signal range and drop out of the network, other cars can join in, connecting vehicles to one another so that a mobile Internet is created. It is a promising approach for future intelligent transportation system (ITS). These networks have no fixed infrastructure and instead rely on the vehicles themselves to provide network functionality. However, due to mobility constraints, driver behavior, and high mobility, VANETs exhibit characteristics that are dramatically different from many generic MANETs.

### Internet Based Mobile Ad-Hoc Network

Internet Based Mobile Ad-hoc Networks are ad hoc networks that link mobile nodes and fixed Internet- gateway nodes. In such type of networks normal ad hoc routing algorithms don't apply directly. Internet Based Mobile Ad-hoc Networks are ad-hoc networks that link mobile nodes and fixed Internet-gateway nodes. In such type of networks normal ad hoc routing algorithms don't apply directly. Wireless networks can generally be classified as wireless fixed networks, and wireless, or mobile ad-hoc networks. MANETs (mobile ad-hoc networks) are based on the idea of establishing a network without taking any support from a centralized structure. By nature these types of networks are suitable for situations where either no fixed infrastructure exists, or to deploy one is not possible.

### Intelligent Vehicular Ad-Hoc Network

Intelligent Vehicular Ad-Hoc Network is a kind of artificial intelligence that helps vehicles to behave in intelligent manners during vehicle-to-vehicle collisions, accidents, drunken driving etc. Intelligent Vehicular Ad-Hoc Networking defines an intelligent way of using Vehicular Networking. In VANET integrates on multiple ad-hoc networking technologies such as Wi-Fi IEEE 802.11, WAVE IEEE 1609, WiMAX, IEEE 802.16, Bluetooth, IRA, and ZigBee for easy, accurate, effective and simple communication between vehicles on dynamic mobility. Effective measures such as media communication between vehicles can be enabled as well methods to track the automotive vehicles are also preferred. InVANET helps in defining safety measures in vehicles, streaming communication between vehicles, infotainment and telematics.

## Wireless Sensor Network

A wireless sensor network (WSN) sometimes called a wireless sensor and actor network (WSAN). WSN are spatially distributed autonomous sensors to monitor physical or environmental conditions, such as temperature, sound, pressure, etc. and to cooperatively pass their data through the network to a main location. The more modern networks are bi- directional, also enabling control of sensor activity. The WSN is built of "nodes" – from a few to several hundreds or even thousands, where each node is connected to one (or sometimes several) sensors. Each such sensor network node has typically several parts: a radio transceiver with an internal antenna or connection to an external antenna, a microcontroller, an electronic circuit for interfacing with the sensors and an energy source, usually a battery. The development of wireless sensor networks was motivated by military applications such as battlefield surveillance; today such networks are used in many industrial and consumer applications, such as industrial process monitoring and control, machine health monitoring, and so on. In computer science and telecommunications, wireless sensor networks are an active research area with numerous workshops and conferences arranged each year, for example IPSN, SenSys, and EWSN.

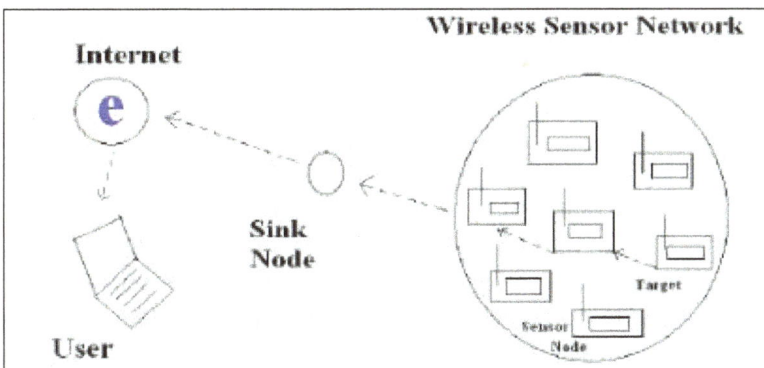

Wireless sensor network2.3 wireless mesh network (wmn).

A wireless mesh network (WMN) is a mesh network created through the connection of wireless access points installed at each network user's locale. A wireless mesh network (WMN) is a communications network made up of radio nodes organized in a mesh topology. Wireless mesh networks often consist of mesh clients, mesh routers and gateways. Well-known examples of WMN are Laptops, cell phone and other wireless devices. The mesh routers forward traffic to and from the gateways which may but need not connect to the Internet. The coverage area of the radio nodes working as a single network is sometimes called a mesh cloud. Access to this mesh cloud is dependent on the radio nodes working in harmony with each other to create a radio network. Wireless mesh architecture is a first step towards providing cost effective and dynamic high-bandwidth networks over a specific coverage area. Wireless mesh infrastructure is, in effect, a network of routers minus the cabling between nodes. Wireless mesh networks can be implemented with various wireless technology including 802.11, 802.15, 802.16, cellular technologies or combinations of more than one type.

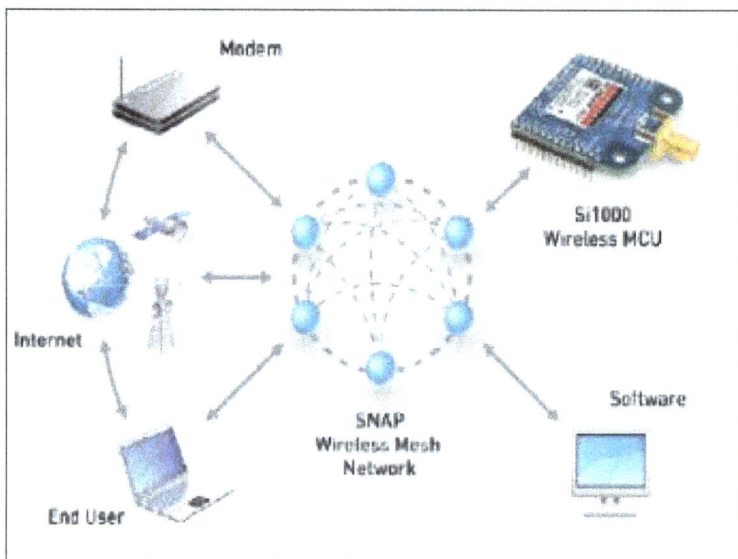

Wireless mesh network.

## Applications of Mobile Ad-Hoc Network

Some of the typical applications include: Collaborative work: For some business environments, the need for collaborative computing might be more important outside office environments than inside and where people do need to have outside meetings to cooperate and exchange information on a given project. Military battlefield: Ad-Hoc networking would allow the military to take advantage of commonplace network technology to maintain an information network between the soldiers, vehicles, and military information head quarter. Local level: Ad-Hoc networks can autonomously link an instant and temporary multimedia network using notebook computers to spread and share information among participants at e.g. conference or classroom. Another appropriate local level application might be in home networks where devices can communicate directly to exchange information. Personal area network and Bluetooth: A personal area network is a short range, localized network where nodes are usually associated with a given person. Short-range MANET such as Bluetooth can simplify the inter communication between various mobile devices such as a laptop, and a mobile phone. Commercial Sector: Ad hoc can be used in emergency/rescue operations for disaster relief efforts, e.g. in fire, flood, or earthquake.

## Applications of Wireless Sensor Network

Area monitoring is a common application of WSNs. In area monitoring, the WSN is deployed over a region where some phenomenon is to be monitored. A military example is the use of sensors to detect enemy intrusion; a civilian example is the geo-fencing of gas or oil pipelines. A disaster alert system that uses WSN and Analytic Network Process (ANP) to predict any possible landslides disasters is proposed. In the authors proposed a flashflood alerting system based on a WSN in a rural area.

## Security Goals

Security is an essential requirement in Wireless ad hoc networks as compared to wired networks. Security is an important issue for ad hoc networks, especially for those security-sensitive applications. Security in wireless network is becoming more and more important while the using of mobile equipments such as cellular phones or laptops is tremendously increasing. Security in MANETs is challenging task and difficult to achieve as there is no central server and base station. In fact, the security hole provided by Ad hoc networking is not only the Ad hoc network itself, but the bridge it provides into other networks.

- Availability: Availability means the assets are accessible to authorized parties at appropriate times. Availability applies both to data and to services. It ensures the survivability of network service despite denial of service attack.

- Confidentiality: Confidentiality ensures that computer-related assets are accessed only by authorized parties. Protection of information which is exchanging through a MANET. It should be protected against any disclosure attack like eavesdropping- unauthorized reading of message.

- Integrity: Integrity means that assets can be modified only by authorized parties or only in authorized way. Integrity assures that a message being transferred is never corrupted.

- Authentication: Authentication is essentially assurance that participants in communication are authenticated and not impersonators. The recourses of network should be accessed by the authenticated nodes.

- Authorization: This property assigns different access rights to different types of users. For example a network management can be performed by network administrator only.

- Non-repudiation: Non-repudiation will facilitate the ability to identify the attackers even after the attack happens. This prevents cheaters from denying their crimes. This ensures that the information originator cannot deny having sent the message.

# Multicasting in Ad-Hoc Networks

Multicasting is the transmission of datagrams to a group of hosts identified by a single destination address. Multicasting is intended for group-oriented computing. There are more and more applications where one-to-many dissemination is necessary. The multicast service is critical in applications characterized by the close collaboration of teams (e.g. rescue patrol, battalion, scientists, etc.) with requirements for audio and video conferencing and sharing of text and images. The use of multicasting within a network has many benefits. Multicasting reduces the communication costs for applications that send the same data to multiple recipients. Instead of sending via multiple unicasts, multicasting minimizes the link bandwidth consumption, sender and router processing, and delivery delay. Maintaining group membership information and building optimal multicast trees is challenging even in wired networks. However, nodes are increasingly mobile.

One particularly challenging environment for multicast is a mobile ad-hoc network (MANET). A MANET consists of a dynamic collection of nodes with sometimes rapidly changing multi-hop topologies that are composed of relatively low-bandwidth wireless links. Since each node has a limited transmission range, not all messages may reach all the intended hosts. To provide communication through the whole network, a source-to-destination path could pass through several intermediate neighbor nodes. Unlike typical wire line routing protocols, ad-hoc routing protocols must address a diverse range of issues. The network topology can change randomly and rapidly, at unpredictable times. Since wireless links generally have lower capacity, congestion is typically the norm rather than the exception. The majority of nodes will rely on batteries, thus routing protocols must limit the amount of control information that is passed between nodes. The majority of applications for the MANET technology are in areas where rapid deployment and dynamic reconfiguration are necessary and the wire line network is not available.

These include military battlefields, emergency search and rescue sites, classrooms, and conventions where participants share information dynamically using their mobile devices. These applications lend themselves well to multicast operation. In addition, within a wireless medium, it is even more crucial to reduce the transmission overhead and power consumption. Multicasting can improve the efficiency of the wireless link when sending multiple copies of messages by exploiting the inherent broadcast property of wireless transmission. However, besides the issues for any ad-hoc routing protocol list-ed above, wireless mobile multicasting faces several key challenges. Multicast group members move, thus precluding the use of a fixed multicast topology. Transient loops may form during tree reconfiguration. As well, tree reconfiguration schemes should be simple to keep channel overhead low. Many multicast routing protocols have been pro-posed for ad-hoc networks, a survey can be found in. Comparing these protocols is typ-ically done based on extensive simulation studies. Bagrodia simulated several multicast routing protocols developed specifically for MANET, some tree-based, some based on a mesh structure.

The reported results show that mesh protocols performed significantly better than the tree protocols in mobile scenarios. Lim and Kim evaluated multicast tree construction and proposed two new flooding methods that can improve the performance of the clas-sic flooding method. Royer and Perkins explored the effect of the radio transmission range on the AODV protocol. They found that larger transmission ranges have many benefits (smaller trees, less frequent link breakages), but also cause more network nodes to be affected by multicast data transmission and reduce the effective bandwidth. They conclude that the transmission range should be adjusted to meet the targeted throughput while minimizing battery power consumption. Within the MANET working group at the IETF, two proposed multicast routing protocols for ad-hoc networks are AODV and ODMRP. To avoid confusion with the unicast functionality of ADOV, we will refer to the multicast operation of AODV as the MAODV protocol. To date, no side-by-side comparison of MAODV and ODMRP has been done. We decided to implement these two widely discussed multicast routing protocols for ad-hoc networks in ns-2.

## Multicast Protocols for Mobile Ad-Hoc Networks

## Multicast Ad-Hoc On-demand Distance Vector Protocol

The MAODV (Multicast Ad-hoc On-Demand Distance Vector) routing protocol discov-ers multicast routes on demand using a broadcast route-discovery mechanism. A mo-bile node originates a Route Request (RREQ) message when it wishes to join a multi-cast group, or when it has data to send to a multicast group but it does not have a route to that group. Only a member of the desired multicast group may respond to a join RREQ. If the RREQ is not a join request, any node with a fresh enough route (based on group sequence number) to the multicast group may respond. If an intermediate node receives a join RREQ for a multicast group of which it is not a member, or if it receives a RREQ and it does not have a route to that group, it rebroadcasts the RREQ to its neighbors.

As the RREQ is broadcast across the network, nodes set up pointers to establish the reverse route in their route tables. A node receiving a RREQ first updates its route ta-ble to record the sequence number and the next hop information for the source node. This reverse route entry may later be used to relay a response back to the source. For join RREQs, an additional entry is added to the multicast route table. This entry is not activated unless the route is selected to be part of the multicast tree. If a node receives a join RREQ for a multicast group, it may reply if it is a member for the multicast group's tree and its recorded sequence number for the multicast group is at least as great as that contained in the RREQ. The responding node updates its route and multicast route tables by placing the requesting node's next hop information in the tables, and then unicasts a Request Response (RREP) back to the source node. As nodes along the path to the source node receive the RREP, they add both a route table and a multicast route table entry for the node from which they received the RREP, thereby creating the forward path.

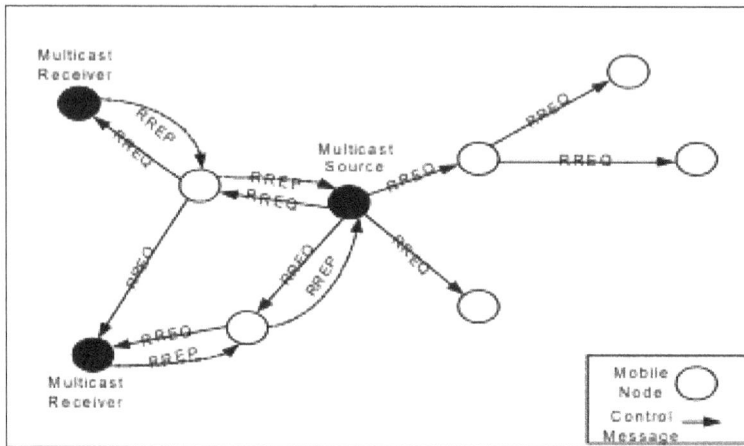

MAODV path discovery.

When a source node broadcasts a RREQ for a multicast group, it often receives more than one reply. The source node keeps the received route with the greatest sequence number and shortest hop count to the nearest member of the multicast tree for a specified period of time, and disregards other routes. At the end of this period, it enables the selected next hop in its multicast route table, and unicasts an activation message (MACT) to this selected next hop. The next hop, on receiving this message, enables the entry for the source node in its multicast route table. If this node is a member of the multicast tree, it does not propagate the message any further. However, if this node is not a member of the multicast tree, it will have received one or more RREPs from its neighbors. It keeps the best next hop for its route to the multicast group, unicasts MACT to that next hop, and enables the corresponding entry in its multicast route table. This process continues until the node that originated the RREP (member of tree) is reached. The activation message ensures that the multicast tree does not have multiple paths to any tree node. Nodes only forward data packets along activated routes in their multicast route tables.

The first member of the multicast group becomes the leader for that group. The multicast group leader is responsible for maintaining the multicast group sequence number and broadcasting this number to the multicast group. This is done through a Group

Hello message. The Group Hello contains extensions that indicate the multicast group IP address and sequence numbers (incremented every Group Hello) of all multicast groups for which the node is the group leader. Nodes use the Group Hello information to update their request table.

Since AODV keeps hard state in its routing table, the protocol has to actively track and react to changes in this tree. If a member terminates its membership with the group, the multicast tree requires pruning. Links in the tree are monitored to detect link breakages. When a link breakage is detected, the node that is further from the multicast group leader (downstream of the break) is responsible for repairing the broken link. If the tree cannot be reconnected, a new leader for the disconnected downstream node is chosen as follows. If the node that initiated the route rebuilding is a multicast group member, it becomes the new multicast group leader. On the other hand, if it was not a group member and has only one next hop for the tree, it prunes itself from the tree by sending its next hop a prune message. This continues until a group member is reached.

Once separate partitions reconnect, a node eventually receives a Group Hello for the multicast group that contains group leader information that differs from the information it already has. If this node is a member of the multicast group, and if it is a member of the partition whose group leader has the lower IP address, it can initiate reconnection of the multicast tree.

## On-demand Multicast Routing Protocol

ODMRP (On-demand Multicast Routing Protocol is mesh based, and uses a forwarding group concept (only a subset of nodes forwards the multicast packets). A soft-state approach is taken in ODMRP to maintain multicast group members. No explicit control message is required to leave the group.

In ODMRP, group membership and multicast routes are established and updated by the source on demand. When a multicast source has packets to send, but no route to the multicast group, it broadcasts a Join-Query control packet to the entire network. This Join-Query packet is periodically broadcast to refresh the membership information and update routes. When an intermediate node receives the Join-Query packet, it stores the source ID and the sequence number in its message cache to detect any potential duplicates. The routing table is updated with the appropriate node ID (i.e. backward learning) from which the message was received for the reverse path back to the source node. If the message is not a duplicate and the Time-To-Live (TTL) is greater than zero, it is rebroadcast.

When the Join-Query packet reaches a multicast receiver, it creates and broadcasts a "Join Reply" to its neighbors. When a node receives a Join Reply, it checks if the next hop node ID of one of the entries matches its own ID. If it does, the node realizes that it is on the path to the source and thus is part of the forwarding group and sets the FG_FLAG (Forwarding

Group Flag). It then broadcasts its own Join Table built upon matched entries. The next hop node ID field is filled by extracting information from its routing table. In this way, each forward group member propagates the Join Reply until it reaches the multicast source via the selected path (shortest). This whole process constructs (or updates) the routes from sources to receivers and builds a mesh of nodes, the forwarding group.

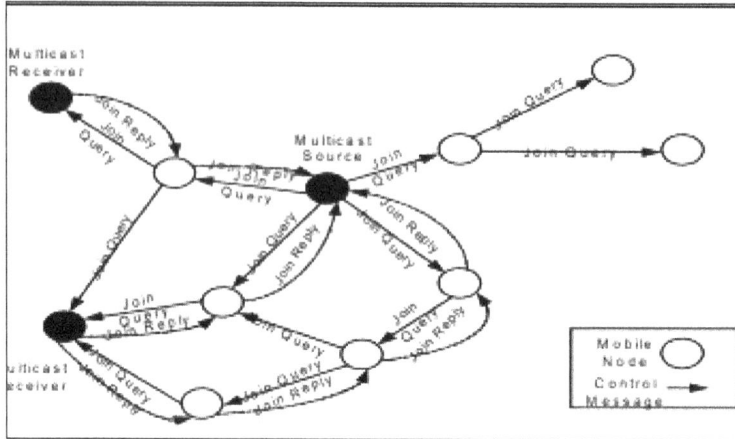

ODMRP mesh creation.

After the forwarding group establishment and route construction process, sources can multicast packets to receivers via selected routes and forwarding groups. While it has data to send, the source periodically sends Join-Query packets to refresh the forwarding group and routes. When receiving the multicast data packet, a node forwards it only when it is not a duplicate and the setting of the FG_FLAG for the multicast group has not expired. This procedure minimizes the traffic overhead and prevents sending packets through stale routes.

In ODMRP, no explicit control packets need to be sent to join or leave the group. If a multicast source wants to leave the group, it simply stops sending Join-Query packets since it does not have any multicast data to send to the group. If a receiver no longer wants to receive from a particular multicast group, it does not send the Join Reply for that group.

Nodes in the forwarding group are demoted to non-forwarding nodes if not refreshed (no Join Tables received) before they timeout.

## Qualitative Comparison of MAODV and ODMRP

The two on-demand protocols share certain salient characteristics. In particular, they both discover multicast routes only in the presence of data packets to be delivered to a multicast destination. Route discovery in either protocol is based on request and reply cycles where multicast route information is stored in all intermediate nodes on the multicast path. However, there are several important differences in the dynamics of the two protocols, which may give rise to significant performance differences.

First, MAODV uses a shared bi-directional multicast tree while ODMRP maintains a mesh topology rooted from each source. In MAODV, the tree is based on hard state and any link breakages force actions to repair the tree. A multicast group leader maintains up to date multicast tree information by sending periodic group hello messages. ODMRP provides alternative paths and a link failure need not trigger the recomputation of the mesh, broken links will time out (soft state). Routes from multicast source to receivers in ODMRP are periodically refreshed by the source. However, a bi-directional tree is more efficient and avoids sending duplicate packets to receivers. Also, depending on the refresh interval in ODMRP, the control overhead from sending route refreshes from every source could result in scalability issues.

Second, ODMRP broadcasts the reply back to the source while MAODV unicasts the reply. By using broadcasts, ODMRP allows for multiple possible paths from the multicast source back to the receiver. Since MAODV unicasts the reply back to the source, if an intermediate node on the path moves away, the reply is lost and the route is lost. However, a broadcasted reply requires intermediate nodes not interested in the multicast group to drop the control packets, resulting in extra processing overhead.

Third, MAODV does not activate a multicast route immediately while ODMRP does (unless mobility prediction is enabled). In MAODV, a potential multicast receiver must wait for a specified time allowing for multiple replies to be received before sending an activation message along the multicast route that it selects.

## Simulation-based Comparison

The performance simulation environment used is based on ns-2, a network simulator that provides support for simulating multi-hop wireless networks complete with physical and IEEE 802.11 MAC layer models.

## Experimental Setup and Performance Metrics

The simulated environment consists of 50 wireless mobile nodes roaming in a 1000 meters x 1000 meters flat space for 900 seconds of simulated time. The radio transmission range is 250 meters. A free space propagation channel is assumed. Group scenario files determine which nodes are receivers or sources and when they join or leave a group. A multicast member node joins the multicast group at the beginning of the simulation (first 30 seconds) and remains as a member throughout the whole simulation. Hence, the simulation experiments do not account for the overhead produced when a multicast member leaves a group. Multicast sources start and stop sending packets in the same fashion (four packets per second, each packet has a constant size of 512 bytes). Each data point represents an average of at least five runs with identical traffic models, but different randomly generated mobility scenarios. For fairness, identical mobility and traffic scenarios are used across the compared protocols. Only one multicast group was used for all the experiments.

Each mobile node moves randomly at a preset average speed according to a "random waypoint model". Here, each node starts its journey from a random location to a random destination with a randomly chosen speed (uniformly distributed between 0 – some maximum speed). Once the destination is reached, another random destination is targeted after a pause. By varying the pause time, the relative speeds of the mobiles are affected. In our experiments the pause time was always set to zero to create a harsher mobility environment. The maximum speeds used were chosen from between 1m/s to 20m/s.

The following metrics were used in comparing the protocol performance. The metrics were derived from ones suggested by the IETF MANET working group for routing/multicast protocol evaluation:

- Packet Delivery Ratio: The ratio of the number of packets actually delivered to the destinations versus the number of data packets supposed to be received. This number presents the effectiveness of a protocol in delivering data to the intended receivers within the network.

- Number of data packets transmitted per data packet delivered: "Data packets transmitted" is the count of every individual transmission of data by each node over the entire network. This count includes transmissions of packets that are eventually dropped and retransmitted by intermediate nodes.

- Number of control packets transmitted per data packet delivered: This measure shows the efficiency overhead in control packets expended in delivering a data packet to an intended receiver.

- Number of control packets and data packets transmitted per data packet delivered: This measure tries to capture a protocol's channel access efficiency, as the cost of channel access is high in contention-based link layers.

To test the protocols, we performed a number of experiments to explore the performance of MAODV and ODMRP with respect to a number of parameters: number of senders, node mobility, and multicast group size.

## Number of Senders

We varied the number of senders in the multicast group in order to evaluate the protocol scalability with respect to source nodes and the resulting effective traffic load. ODMRP is over 53% more effective than MAODV in data delivery ratio as the number of senders is increased from one to twenty. In terms of packet transmission ratio though, at twenty senders, MAODV sends 75% fewer packets for each data packet delivered than ODMRP. As well, MAODV sends 59% fewer control overhead packets than ODMRP for each data packet delivered as the number of senders reaches twenty. For both control and data transmissions, MAODV sends 90% less packets than ODMRP for every packet delivered as the number of senders reaches twenty.

We observed that ODMRP in particular does not scale well for packet delivery ratio as the number of senders increases along with the effective traffic load. In ODMRP, every source node will periodically send out route requests through the network. When the number of source nodes becomes larger, the effect of this causes congestion in the network and the data delivery ratio drops significantly. MAODV, on the other hand, maintains only one group leader for the multicast group that will send periodic Group Hellos through the network. In this manner, it is more scalable than ODMRP.

## Node Mobility

We varied the mobility to evaluate the ability of the protocols to deal with route changes. ODMRP is over 104% more effective than AODV in data delivery ratio as the maximum node speed is increased from 1m/s to 20m/s. In terms of packet transmission ratio, ODMRP sends 40% less packets for each data packet delivered at high mobility (>15m/s). As well, for control overhead, ODMRP decreases by up to 74% less than MAODV for each data packet delivered as the mobility reaches 20m/s. For control and data transmissions, ODMRP sends 48% less packets than MAODV for every packet delivered. We see that ODMRP is generally unaffected by increases in mobility, while MAODV is more sensitive to changes in mobility. The mesh topology of ODMRP allows for alternative paths thus making it more robust than MAODV. MAODV relies on a single path on its multicast tree, and must react to broken links, by initiating repairs.

## Multicast Group Size

For the third set of simulations, we varied the number of members in the multicast group in order to evaluate the protocol scalability with respect to multicast group size. ODMRP is 270% to 20% more effective than MAODV in data delivery ratio as the number of multicast group members is increased from ten to fifty. In terms of packet transmission ratio MAODV sends up to 48% less packets for each data packet delivered. As well, for control and data transmissions MAODV decreases by up to 46% less than ODMRP for each data packet delivered.

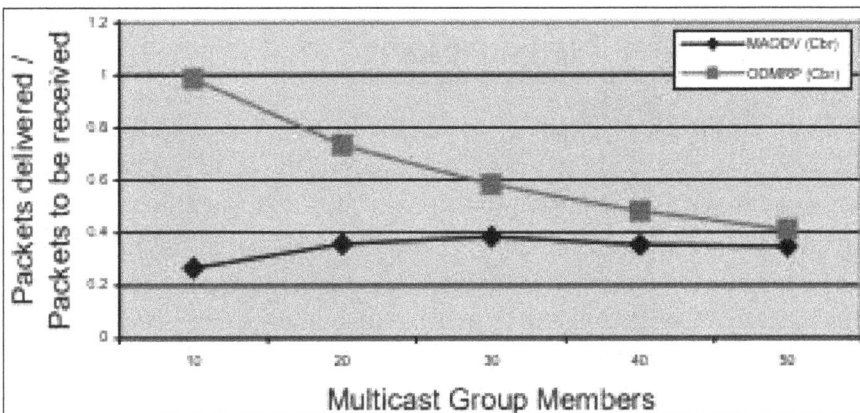

Data delivery ratio as a function of multicast group size.

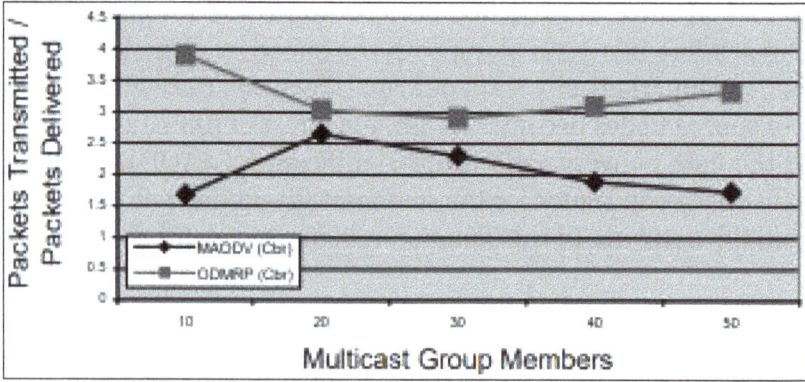

Packet transmission ratio as a function of multicast group size.

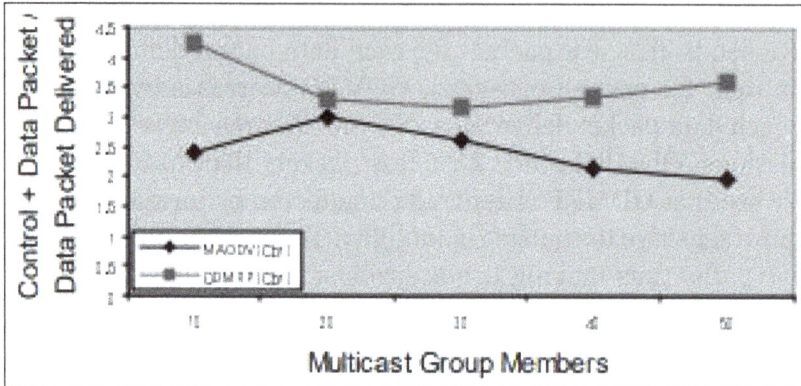

Control and data transmissions per data packet delivered vs. group size.

ODMRP does not scale well with multicast group size. There is a drastic decline in packet delivery ratio as the multicast group increases to fifty members. This can be attributed to collisions that occur from the frequent broadcasts through the network. Despite the poor data delivery ratio, we see that MAODV scales better in terms of over-all control and data transmissions for every packet delivered.

# Mobile Ad-Hoc Networks

Many new applications are resulted from progress in the internet discipline because of wireless network tech-nologies. For research and development of wireless network, one of the most auspicious arenas is Mobile Ad-Hoc Network (MANET). Wireless ad-hoc network is becoming one of the most animated and dynamic field of communication and networks because of fame of movable device and wireless networks that has in-creased significantly in recent years. A mobile ad-hoc network is formed by collecting portable devices like laptops, smart phones, sensors, etc. that communicate through wireless links with one another. These devices collaborate with each other to offer the Many new applications are resulted from progress in the internet discipline because of wireless network technologies. For research and development of wireless network, one of the most auspicious arenas is Mobile Ad- Hoc Network (MANET). Wireless ad-hoc network is becoming one of the most animated and dynamic field of communication and networks because of fame of movable device and wireless networks that has increased significantly in recent years. A mobile ad-hoc network is formed by collecting portable devices like laptops, smart phones, sensors, etc. that communicate through wireless links with one another. These devices collaborate with each other to offer the essential network functions in the nonappearance of immovable organization in a distributed manner. This type of network creates the way for various innovative and stimulating applications by functioning as an independent network or with multiple points of connection to cellular networks or the Internet.

Routing of packets to destination is done by the cooperation of nodes of a MANET. The sending and receiving devices may be situated at a much higher distance as compared to transmission radius R, however, each network node can communicate only with nodes placed within its broadcast radius R. All the nodes in a multi- hop wireless ad-hoc network collaborate with one another to create a network in the absence of infrastructure such as access point or base station.

In order to permit transmission among devices beyond the transmission range in MANET, the mobile devices require advancing data packets for one another. The network devices can move freely and autonomously in any route. The nodes can detach and attach to the network haphazardly. Thus variations in link states of the node with other nodes are experienced by a node regularly. Challenges for routing protocols operating in MANET are eventually increased the movement in the ad-hoc network, changes in link states and other characteristics of wireless transmission such as attenuation, multipath propagation, interference etc. The challenges are boosted by the numerous sorts of nodes of restricted processing power and competences that may join the network.

Ultimate aim and objective of this research work is to analytically review routing protocols of MANET, simulate DSR, TORA and OLSR routing protocols by using simulator and compare the results under different scenarios like with Nodes Density of 25, 50 and 75 nodes, evaluate and analyze these routing protocols under the various environments by using some parameters like WLAN delay, WLAN throughput, WLAN network load, FTP traffic sent and received both by the nodes and server, routing traffic sent and received.

MANET is basically an organization less network of transportable devices having wireless communication capabilities that can join together at any time and at any place dynamically. In this type of network mobile hosts, sometimes, simultaneously acting as a router, are connected to one another by wireless links and they can easily move randomly hence network topology dynamically change so this makes an autonomous system of mobile nodes having no base station. In MANET each node has limited transmission range so packets are forwarded from any initiating node to any end point node in a network with the help of multiple hopes.

## Wireless Networks

Wireless networks not only offer connection flexibility among users at different places they also help in the ex-tension of the network to any building or area without a phys-ical-wired connection. Such connections are of two types; Substructure networks and Ad-Hoc networks.

Access Point (AP), in Infrastructure wireless networks, denotes a central controller for each device. The net-work can be joined through access point by any node. In order to make the route ready when it is needed, the access point arranges the linking among the Basic Set Services (BSSs). Still, there is a disadvantage of using an organization of network that is the big overhead of preserving the routing tables.

Structure less or Ad-Hoc networks lack firm topology or a central controlling point, that's why, transfer and receipt of data-packets is more complex as compared to struc-tured networks. Wireless networks can be classified as single hop and multi hop as well as infrastructure based and ad hoc based. In single hop wireless networks base station (BS) and wireless devices communicated directly with electromagnetic waves to each other. In multi hop wireless networks wireless devices indirectly communicate with the base station by sending data from one device to other and so on.

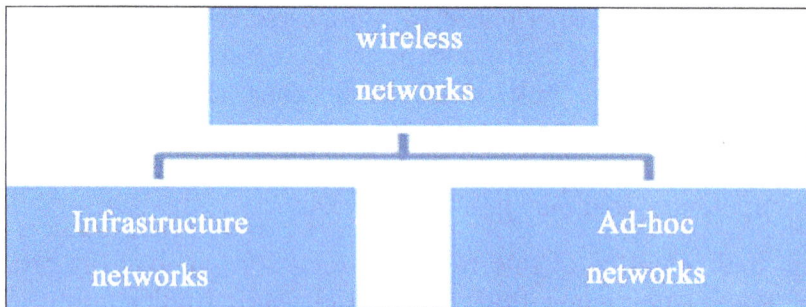

Classifications of wireless networks.

## Characteristics of MANET

Autonomous and infrastructure-less: MANET is independent of conventional structure or central management. Each device plays the role of an independent router and gener-ates independent data because it functions in dispersed P2P style. Fault detection and management becomes difficult as network administration has to be scattered crosswise various nodes.

Multi-hop routing: Every node plays the role of router and forward packets for infor-mation distribution among portable hosts. No default router is available. Dynamic topologies: Because of arbitrary movement of nodes, the network topography (which is classically multi-hop) changes regularly and randomly. This results in changes in routes, common network sub divisions and perhaps data-packet losses.

Unconventionality in link and node abilities: It is possible for every device to be armed with one or more radio interfaces with each one having changing sending/receiving competences. These may also function transversely various frequency groups. Irregular links may be resulted due to these heterogeneous node radio abilities. In addition to this, processing capability may also vary due to different software/hardware configuration of mobile nodes. Scheming network protocols and algorithms for this diverse network can be complicated, demanding dynamic adjustments to the altering situations (power and channel conditions, traffic load/distribution variations, congestion, etc.). Energy controlled process. Each movable device having restricted battery power supply, processing power is affected. As a result facilities and applications provided by every device are also limited. Additional energy is required because every node has to act as system and router. Additional power is required for packer forwarding to other nodes. So this develops to be a greater problem in mobile ad hoc networks.

Network scalability: Scalability is serious issue to the successful implementation of these networks. Currently, popular network management algorithms could only support static or comparatively minor wireless networks. Numerous mobile ad hoc network applications include large networks huge number of nodes, as found for instance, in sensor networks and tactical networks. The implementation of such a network presents several issues that are needed to be resolved. These challenges include: addressing, routing, location management, configuration management, interoperability, security, high capacity wireless technologies, etc.

## Benefits of MANET

- Highly suitable network in such circumstances where fixed infrastructure is too much costly, untrustworthy, not trusted and due to unavailability of such a network.

- Quickly installation with least possible user intervention.

- Detailed planning and installation of base stations is not required.

- Ad hoc networks can be attached to the WWW or Internet, thereby incorporating many different devices and making possible for other users to use available services.

- Capacity, range and energy arguments promote their use in tandem with existing cellular infrastructures as they can extend coverage and interconnectivity.

- MANET also fitted to use the future 4G architecture and their services, aims to provide ubiquitous computer environments that support users in completing their tasks, accessing information and communicating anywhere, anytime and from any device.

## MANET Environment Variations

The different MANET setting differences are programmed taking its dynamic topology in to consideration.

In MANETs all the nodes have duplicate abilities and responsibilities, which are labeled as symmetric environment. MANET containing wireless mobile nodes that interconnects without integrated control or recognized structure. These nodes are within each other's radio range can interconnect directly, while expanse nodes depend on their adjoining nodes to forward the packets. In MANETs these node can be a router or a host. In MANET environment, these nodes are allowed to leave or join the system at any time, ensuring a highly energetic network environment paralleled to supported network.

The Irregular Competences in MANETs comprise transmission series and radios series which may change. Speed of movement, battery life and processing capacity will be dissimilar at different nodes. Irregular Responsibilities contain that some of the nodes may track packets in the network or some of the nodes may perform as leaders for the neighboring nodes such as cluster head. Traffic features may differ in diverse ad hoc networks like timeliness constraints, bit rate reliability necessities, unicast, multicast or geocast, content-based addressing or host-based addressing or capability-based addressing. MANETs may co-operate and also co-exist with a setup based network. Mobility arrangements may be different as people sitting at any airport lounge, taxi cabins, military actions and private area networks. The action of a mobile ad hoc network is reliant on the node mobility design as well as data traffic patterns, topology and radio intervention. Mobility features including speed, direction of movement, predictability, design of movement, consistency of mobility features among different nodes.

## MANET Challenges

The subsequent list of issues indicates the inadequacies and restrictions that have to be overwhelmed in a MANET environment:

- Restricted wireless transmission range: The radio group will be restricted in the wireless networks and as a result data amounts it can provide much slighter than what a bound network can provide. This involves routing procedures of wireless networks must be use bandwidth in ideal way. This can be achieved through protecting the overhead as minimum as conceivable. The restricted transmission range also enforces restraint on routing procedures for sustaining the topographical information. Particularly in MANETs because of regular variations in topology, preserving the topological data for every node includes more controllers overhead which results in additional bandwidth depletion.

- Time-varying wireless link characteristics: Wireless channel is liable to a range of broadcast disorders such as path harm, declining, intervention and obstruction. These features resist the series, data rate, and consistency of these cordless transmissions. The range of which these features disturb the transmission that rest on atmospheric situations and flexibility of receiver and transmitter. Even two dissimilar key restraints, Nyquist's and Shannon's theorems that rule over capability to communicate the information at diverse data degrees can be measured.

- Broadcast nature of the wireless medium: The broadcast nature of the radio channel, such as transmissions prepared by a device is established by all devices that are in its straight transmission covering area. When a device receives data, no other device in its neighborhood, apart from the sender, must transfer. A device can acquire access to the mutual medium when its communications cannot disturb any constant session. Meanwhile several devices may resist for medium contemporarily, chance of data-packet crashes is very tall in wireless networks. Even the network is liable to concealed terminal issue and transmits storms. Concealed terminal issue mentions to the smash of data-packets at a receipt device because of immediate transmission of the nodes which are outside the straight communication series of the transmitter, but are inside the communication series of the receiver.

- Packet losses due to transmission errors: Ad hoc wireless networks practices very advanced packet damage due to reasons such as extraordinary bit error rate (BER) in the wireless channel, enlarged crashes because of the existence of unseen terminals, occurrence of interventions, position reliant controversy, single directional associations, regular pathway breakages due to device movements, and the integral declining characteristics of the wireless passage.

- Mobility-induced route changes: The system topography in ad hoc wireless network is extremely active because of node movement; as a result, a constant meeting undergoes numerous pathway breakages. Such position often results in regular path alterations. So flexibility administration is massive investigation theme in ad hoc networks.

- Mobility-induced packet losses: Communication contacts in an ad hoc network are insecure such that consecutively conservative procedures for MANETs over a great damage frequency will suffer from performance deprivation. Though, with large frequency of inaccuracy, it is problematic to supply a data-packet to its target.

- Battery constraints: It is due to restricted resources that arrange main limitation on the mobile devices in an ad hoc network. Nodes which are contained in such network have restrictions on the supremacy foundation in order to preserve movability, dimension and capacity of the node. Due to accumulation of power and the processing capacity make the nodes heavyweight and less portable. Consequently only MANET devices have to use this resource.

- Potentially frequent network partitions: Casually stirring nodes in an ad-hoc network may result in network panels. Certain cases involve middle nodes to be extremely effected by such separation.

- Ease of snooping on wireless transmissions (security issues): Wireless passage being employed for ad hoc networks transmitted in natural surroundings. It is also shared by all devices in the network. Transmission of data through a device is acknowledged by all devices inside straight communication series. So invader is certain to sneak data/information which is communicated within network. The conditions of secrecy could be disrupted if enemy is capable in inferring data assembled by snooping.

- Routing: In MANETs routing is an important challenge for the performance degradation due to unicasting, multicasting and geocasting demands by the network nodes in contrast to single hope wireless networks. It's because of rapid change in network topology and with different mobility speeds.

- Quality of Service: In MANETs quality of service is an important challenge for the differed kind of quality level demands by the network nodes. Its becomes very difficult to fulfill the different levels or priority demands related to quality of service so these network required best control of QoS specially in case of multimedia.

- Security: In MANET, security is one the important challenge due to its wireless environment. The data of users from one node to another node must be transferred safely and completely. The least privilege principle can also enhance the security of MANET systems as proposed for organizations. Moreover, there are hybrid models are also available that are offering benefits of two access control models with implementations.

## Applications of MANET

Some distinctive MANET applications include:

- Military field: Ad-Hoc networking can permit army to exploit benefit of conventional network expertise for preserving any info network among vehicles, armed forces, and headquarters of information.

- Cooperative work: To facilitate the commercial settings, necessity for concerted computing is very significant external to office atmosphere and surroundings as compared to inner environment. People want getting outside meetings for exchanging the information plus cooperating with each other regarding any assigned task.

- Confined level: Ad-Hoc networks are able to freely associate with immediate, in addition momentary hyper-media network by means of laptop computers for sharing the info with all the contestants' e.g. classroom and conference. Additional valid and confined level application may be in domestic network where these devices can interconnect straight in exchanging the information.

- PAN and Bluetooth: A PAN is localized and tiny range network whose devices are generally belong to a specified individual. Limited-range MANET such as Bluetooth can make simpler the exchange among several portable devices like a laptop, and a cell phone.

- Business Sector: Ad-hoc network could be used for rescuing and emergency processes for adversity assistance struggles, for instance, in flood, fire or earthquake. Emergency saving procedures should take place where damaged and non-existing transmissions structure and quick preparation of a transmission network is required.

- Sensor Networks: managing home appliances with MANETs in both the case like nearby and distantly. Tracking of objects like creatures. Weather sensing related activities.

- Backup Services: liberation operations, tragedy recovery, diagnosis or status or record handing in hospitals, replacement of stationary infrastructure.

- Educational sector: arrangement of communications facilities for computer-generated conference rooms or classrooms or laboratories.

## Routing in Mobile Ad-Hoc Networks

The absence of fixed infrastructure in a MANET poses several types of challenges. The biggest challenge among them is routing. Routing is the process of selecting paths in a network along which to send data packets. An ad hoc routing protocol is a convention, or standard, that controls how nodes decide which way to route packets between computing devices in a mobile ad-hoc network. In ad hoc networks, nodes do not start out familiar with the topology of their networks; instead, they have to discover it. The basic idea is that a new node may announce its presence and should listen for announcements broadcast by its neighbors. Each node learns about nearby nodes and how to reach them, and may announce that it can reach them too. The routing process usually directs forwarding on the basis of routing tables which maintain a record of the routes to various network destinations. Thus, constructing routing tables, which are held in the router's memory, is very important for efficient routing.

### Routing Protocols for MANET

The growth of laptops and 802.11/Wi-Fi wireless networking has made MANETs a popular research topic since the 1990s. The proposed solutions for routing protocols could be grouped in three categories: proactive (or table-driven), reactive (or on-demand), and hybrid protocols. Even the reactive protocols have become the main stream for MANET routing. In this chapter, we introduce some popular routing protocols in each of the three categories and for IPv6 networks.

## Applications for MANET

Ad hoc networks are suited for use in situations where infrastructure is either not available or not trusted, such as a communication network for military soldiers in a field, a mobile network of laptop computers in a conference or campus setting, temporary offices in a campaign headquarters, wireless sensor networks for biological research, mobile social networks such as Facebook, MySpace and Twitter, and mobile mesh networks for Wi-Fi devices.

## Proactive Routing Protocols

Every proactive routing protocol usually needs to maintain accurate information in their routing tables. It attempts to continuously evaluate all of the routes within a network. This means the protocol maintains fresh lists of destinations and their routes by periodically distributing routing tables throughout the network. So that when a packet needs to be forwarded, a route is already known and can be used immediately. Once the routing tables are setup, then data (packets) transmissions will be as fast and easy as in the tradition wired networks. Unfortunately, it is a big overhead to maintain routing tables in the mobile ad hoc network environment. Therefore, the proactive routing protocols have the following common disadvantages: 1. Respective amount of data for maintaining routing information. 2. Slow reaction on restructuring network and failures of individual nodes. Proactive routing protocols became less popular after more and more reactive routing protocols were introduced. Here, we introduce three popular proactive routing protocols – DSDV, WRP and OLSR. Besides the three popular protocols, there are many other proactive routing protocols for MNAET, such as CGSR, HSR, MMRP and so on.

## Destination-Sequenced Distance Vector (DSDV)

Destination-Sequenced Distance-Vector Routing (DSDV) is a table-driven routing scheme for ad hoc mobile networks based on the Bellman-Ford algorithm. It was developed by C. Perkins and P. Bhagwat in 1994. The main contribution of the algorithm was to solve the routing loop problem. Each entry in the routing table contains a sequence number. If a link presents the sequence numbers are even generally, otherwise an odd number is used. The number is generated by the destination, and the emitter needs to send out the next update with this number. Routing information is distributed between nodes by sending full dumps infrequently and smaller incremental updates more frequently.

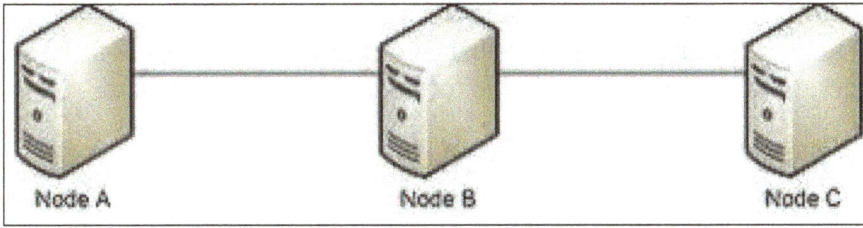

For example the routing table of Node A in the above network is:

| Destination | Next Hop | Number of Hops | Sequence Number | Install Time |
|---|---|---|---|---|
| A | A | 0 | A46 | 001000 |
| B | B | 1 | B36 | 001200 |
| C | B | 2 | C28 | 0015000 |

Naturally the table contains description of all possible paths reachable by node A, along with the next hop, number of hops, sequence number and install time.

## Selection of Route

If a router receives new information, then it uses the latest sequence number. If the sequence number is the same as the one already in the table, the route with the better metric is used. Stale entries are those entries that have not been updated for a while. Such entries as well as the routes using those nodes as next hops are deleted. Then new destination comes. This is how it works.

## Influence

Since no formal specification of this algorithm is present, there is no commercial implementation of this algorithm. But some other protocols have used similar techniques. The best-known sequenced distance vector protocol is AODV, which, by virtue of being a reactive protocol, can use simpler sequencing heuristics. Besides, Babel is a distance-vector routing protocol for IPv4 and IPv6 with fast convergence properties. It was designed to make DSDV more robust, more efficient and more widely applicable for both wireless mesh networks and classical wired networks while staying within the framework of proactive protocols.

## Advantages

DSDV was one of the early algorithms available. It is quite suitable for creating ad hoc networks with small number of nodes.

## Disadvantages

DSDV requires a regular update of its routing tables, which uses up battery power and a small amount of bandwidth even when the network is idle. Also, whenever the topology of the network changes, a new sequence number is necessary before the network re-converges; thus, DSDV is not suitable for highly dynamic networks.

## Wireless Routing Protocol (WRP)

The Wireless Routing Protocol (WRP) is a proactive unicast routing protocol for MANETs. WRP uses an enhanced version of the distance-vector routing protocol, which uses the Bellman-Ford algorithm to calculate paths. Because of the mobile nature of the nodes within the MANET, the protocol introduces mechanisms which reduce route loops and ensure reliable message exchanges. The wireless routing protocol (WRP), similar to DSDV, inherits the properties of the distributed Bellman-Ford algorithm. To solve the count-to-infinity problem and to enable faster convergence, it employs a unique method of maintaining information regarding the shortest path to every destination node and the penultimate hop node on the path to every destination node in the network. Since WRP, like DSDV, maintains an up-to-date view of the network, every node has a readily available route to every destination node in the network. It differs from DSDV in table maintenance and in the update procedures. While DSDV maintains only one topology table, WRP uses a set of tables to maintain more accurate information. The tables that are maintained by a node are the following: distance table (DT), routing table (RT), link cost table (LCT), and a message retransmission list (MRL).

## Distance Table

The DT contains the network view of the neighbors of a node. It contains a matrix where each element contains the distance and the penultimate node reported by a neighbor for a particular destination.

## Routing Table

The RT contains the up-to-date view of the network for all known destinations. It keeps the shortest distance, the predecessor node (penultimate node), the successor node (the next node to reach the destination), and a flag indicating the status of the path. The path status may be a simple path (correct), or a loop (error), or the destination node not marked (null, invalid route). Note, storing the previous and successive nodes assists in detecting loops and avoiding the counting-to-infinity problem a shortcoming of Distance Vector Routing.

## Link Cost Table

The LCT contains the cost (e.g., the number of hops to reach the destination) of relaying messages through each link. The cost of a broken link is infinity. It also contains the number of update periods (intervals between two successive periodic updates) passed since the last successful update was received from that link. This is used to detect link breaks. The LCT maintains the cost of the link to its nearest neighbors (nodes within direct transmission range), and the number of timeouts since successfully receiving a message from the neighbor. Nodes periodically exchange routing tables with their neighbors via update messages, or whenever the link cost table changes.

## Message Retransmission List

The MRL contains an entry for every update message that is to be retransmitted and maintains a counter for each entry. This counter is decremented after every retransmission of an update message. Each update message contains a list of updates. A node also marks each node in the RT that has to acknowledge the update message it transmitted. Once the counter reaches zero, the entries in the update message for which no acknowledgments have been received are to be retransmitted and the update message is deleted. Thus, a node detects a link break by the number of update periods missed since the last successful transmission. After receiving an update message, a node not only updates the distance for transmission neighbors but also checks the other neighbors' distance, hence convergence is much faster than DSDV. The MRL maintains a list of which neighbors are yet to acknowledge an update message, so they can be retransmitted if necessary. If there is no change in the routing table, a node is required to transmit a "hello" message to affirm its connectivity. When an update message is received, a node updates its distance table and reassesses the best route paths. It also carries out a consistency check with its neighbors, to help eliminate loops and speed up convergence.

## Advantages

WRP has the same advantage as that of DSDV. In addition, it has faster convergence and involves fewer table updates.

## Disadvantages

The complexity of maintenance of multiple tables demands a larger memory and greater processing power from nodes in the wireless ad hoc network. At high mobility, the control overhead involved in updating table entries is almost the same as that of DSDV and hence is not suitable for a highly dynamic and for a very large ad hoc wireless network as it suffers from limited scalability.

## Optimized Link State Routing

The Optimized Link State Routing Protocol (OLSR) is an IP routing protocol optimized for mobile ad-hoc networks, which can also be used on other wireless ad-hoc networks. OLSR is a proactive link-state routing protocol, which uses Hello and Topology Control (TC) messages to discover and then disseminate link state information throughout the mobile ad- hoc network. Individual nodes use this topology information to compute next hop destinations for all nodes in the network using shortest hop forwarding paths.

## Features Specific to OLSR

Link-state routing protocols such as OSPF and IS-IS elect a designated router on every link to perform flooding of topology information. In wireless ad-hoc networks, there is different notion of a link, packets can go out the same interface; hence, a different approach is needed in order to optimize the flooding process. Using Hello messages the OLSR protocol at each node discovers 2-hop neighbor information and performs a distributed election of a set of multipoint relays (MPRs). Nodes select MPRs such that there is a path to each of its 2-hop neighbors via a node selected as an MPR. These MPR nodes then forward TC messages that contain the MPR selectors. This functioning of MPRs makes OLSR unique from other link state routing protocols in a few different ways: The forwarding path for TC messages is not shared among all nodes but varies depending on the source, only a subset of nodes source link state information, not all links of a node are advertised but only those that represent MPR selections.

Since link-state routing requires the topology database to be synchronized across the network, OSPF (Open Shortest Path First) and IS-IS (Intermediate System to Intermediate System) perform topology flooding using a reliable algorithm. Such an algorithm is very difficult to design for ad-hoc wireless networks, so OLSR doesn't bother with reliability; it simply floods topology data often enough to make sure that the database does not remain unsynchronized for extended periods of time.

## Messages used in OLSR

OLSR uses the "Hello" messages to find its one hop neighbors and its two hop neighbors through their responses. The sender can then select its multipoint relays (MPR) OLSR uses the "Hello" messages to find its one hop neighbors and its two hop neighbors through their responses. The sender can then select its multipoint relays (MPR) based on the one hop node that offers the best routes to the two hop nodes. Each node has also an MPR selector set, which enumerates nodes that have selected it as an MPR node. OLSR uses Topology Control (TC) messages along with MPR forwarding to disseminate neighbor information throughout the network. Host and Network Association (HNA) messages are used by OLSR to disseminate network route advertisements in the same way TC messages advertise host routes. Below are the formats of Topology and Hello Control messages.

## Topology Control Message

| o(bits 0-9) | | | 2(bits 20-29) | | | | | | | | | | 1(bits 10-19) | | | | 3 | |
|---|---|---|---|---|---|---|---|---|---|---|---|---|---|---|---|---|---|---|
| 0 | 1 | ....... | 9 | 0 | 1 | 2 | 3 | 4 | 5 | 6 | 7 | 8 | 9 | 0 | 1 | ...... | 9 | 0 | 1 |
| ANSN | | | | | | | | | Reserved | | | | | | | | |
| Advertised Neighbor Main Address | | | | | | | | | | | | | | | | | |
| Advertised Neighbor Main Address | | | | | | | | | | | | | | | | | |

Each row has 32 bits.

# Hello Control Message

| 0(bits 0-9) | | | | 1(bits 10-19) | | | | | | | | | | 2(bits 20-29) | | | | | | 3 | |
|---|---|---|---|---|---|---|---|---|---|---|---|---|---|---|---|---|---|---|---|---|---|
| 0 | 1 | ....... | 9 | 0 | 1 | 2 | 3 | 4 | 5 | 6 | 7 | 8 | 9 | 0 | 1 | 3 | 4 | ... | 9 | 0 | 1 |
| Reserved | | | | | | | | | | Hitme | | | | Willingness | | | | | | | |
| Link Code | | | | Reserved | | | | | | Link Message Size | | | | | | | | | | | |
| ANSN | | | | | | | | | | Reserved | | | | | | | | | | | |
| Advertised Neighbor Main Address | | | | | | | | | | | | | | | | | | | | | |
| Advertised Neighbor Main Address | | | | | | | | | | | | | | | | | | | | | |
| .... | | | | | | | | | | | | | | | | | | | | | |
| .... | | | | | | | | | | | | | | | | | | | | | |
| Link Code | | | | Reserved | | | | | | Link Message Size | | | | | | | | | | | |
| Advertised Neighbor Main Address | | | | | | | | | | | | | | | | | | | | | |
| Advertised Neighbor Main Address | | | | | | | | | | | | | | | | | | | | | |

## Advantages

Being a proactive protocol, routes to all destinations within the network are known and maintained before use. Having the routes available within the standard routing table can be useful for some systems and network applications as there is no route discovery delay associated with finding a new route. The routing overhead generated, while generally greater than that of a reactive protocol, does not increase with the number of routes being used. Default and network routes can be injected into the system by HNA (Host and Network Association) messages allowing for connection to the internet or other networks within the OLSR MANET cloud. Network routes using reactive protocols do not currently execute well. Timeout values and validity information is contained within the messages conveying information allowing for differing timer values to be used at differing nodes.

## Disadvantages

The original definition of OLSR does not include any provisions for sensing of link quality; it simply assumes that a link is up if a number of hello packets have been received recently. This assumes that links are bi-modal (either working or failed), which is not necessarily the case on wireless networks, where links often exhibit intermediate rates of packet loss.

Implementations such as the open source OLSRD (OLSR Daemon, commonly used on Linux-based mesh routers) have been extended (as of v. 0.4.8) with link quality sensing. Being a proactive protocol, OLSR uses power and network resources in order to propagate data about possibly unused routes. While this is not a problem for wired access points, and laptops, it makes OLSR unsuitable for sensor networks that try to sleep most of the time. For small scale wired access points with low CPU power, the open source OLSRD project showed that large scale mesh networks can run with OLSRD on thousands of nodes with very little CPU power on 200 MHz embedded devices.

Being a link-state protocol, OLSR requires a reasonably large amount of bandwidth and CPU power to compute optimal paths in the network. In the typical networks where OLSR is used (which rarely exceed a few hundreds of nodes), this does not appear to be a problem. By only using MPRs to flood topology information, OLSR removes some of the redundancy of the flooding process, which may be a problem in networks with moderate to large packet loss rates however the MPR mechanism is self-pruning (which means that in case of packet losses, some nodes that would not have retransmitted a packet may do so).

## OLSR Version 2

OLSRv2 is currently being developed within the IETF. It maintains many of the key features of the original including MPR selection and dissemination. Key differences are the flexibility and modular design using shared components: packet format, and neighborhood discovery protocol (NHDP). These components are being designed to be common among next generation IETF MANET protocols. Differences in the handling of multiple address and interface enabled nodes is also present between OLSR and OLSRv2.

## Reactive Routing Protocols

In bandwidth-starved and power-starved environments, it is interesting to keep the network silent when there is no traffic to be routed. Reactive routing protocols do not maintain routes, but build them on demand. A reactive protocol finds a route on demand by flooding the network with Route Request packets. These protocols have the following advantages:

- No big overhead for global routing table maintenance as in proactive protocols.

- Quick reaction for network restructure and node failure.

Even reactive protocols have become the main stream for MANET routing, they still have the following main disadvantages:

- High latency time in route finding.

- Excessive flooding can lead to network clogging.

There are many reactive routing protocols for MANET. We only introduce three popular (AODV, DSR and DYMO) and one new (ODCR) protocols.

## Ad hoc On-demand Distance Vector

Ad hoc On-Demand Distance Vector (AODV) Routing is a routing protocol for mobile ad hoc networks (MANETs) and other wireless ad-hoc networks. It is jointly developed in Nokia Research Center, University of California, Santa Barbara and

University of Cincinnati by C. Perkins, E. Belding-Royer and S. Das. AODV is capable of both unicast and multicast routing. It is a reactive routing protocol, meaning that it establishes a route to a destination only on demand. In contrast, the most common routing protocols of the Internet are proactive, meaning they find routing paths independently of the usage of the paths. AODV is, as the name indicates, a distance-vector routing protocol. AODV avoids the counting-to- infinity problem of other distance-vector protocols by using sequence numbers on route updates, a technique pioneered by DSDV.

In AODV, the network is silent until a connection is needed. At that point the network node that needs a connection broadcasts a request for connection. Other AODV nodes forward this message, and record the node that they heard it from, creating an explosion of temporary routes back to the needy node. When a node receives such a message and already has a route to the desired node, it sends a message backwards through a temporary route to the requesting node. The needy node then begins using the route that has the least number of hops through other nodes. Unused entries in the routing tables are recycled after a time. When a link fails, a routing error is passed back to a transmitting node, and the process repeats.

Much of the complexity of the protocol is to lower the number of messages to conserve the capacity of the network. For example, each request for a route has a sequence number. Nodes use this sequence number so that they do not repeat route requests that they have already passed on. Another such feature is that the route requests have a "time to live" number that limits how many times they can be retransmitted. The third feature is that if a route request fails, another route request may not be sent until twice as much time has passed as the timeout of the previous route request.

## Technical Description

The AODV Routing protocol uses an on-demand approach for finding routes, that is, a route is established only when it is required by a source node for transmitting data packets. It employs destination sequence numbers to identify the most recent path. The major difference between AODV and Dynamic Source Routing (DSR) is that DSR uses source routing in which a data packet carries the complete path to be traversed; however, in AODV, the source node and the intermediate nodes store the next-hop information corresponding to each flow for data packet transmission.

In an on-demand routing protocol, the source node floods the RouteRequest packet in the network when a route is not available for the desired destination. It may obtain multiple routes to different destinations from a single RouteRequest. The major difference between AODV and other on-demand routing protocols is that it uses a destination sequence number (DestSeqNum) to determine an up-to- date path to the destination. A node updates its path information only if the DestSeqNum of the current packet received is greater than the last DestSeqNum stored at the node.

A RouteRequest carries the source identifier (SrcID), the destination identifier (DestID), the source sequence number (SrcSeqNum), the destination sequence number (DestSeqNum), the broadcast identifier (BcastID), and the time to live (TTL) field. DestSeqNum indicates the freshness of the route that is accepted by the source. When an intermediate node receives a RouteRequest, it either forwards it or prepares a RouteReply if it has a valid route to the destination. The validity of a route at the intermediate node is determined by comparing the sequence number at the intermediate node with the destination sequence number in the RouteRequest packet. If a RouteRequest is received multiple times, which is indicated by the BcastID-SrcID pair, the duplicate copies are discarded. All intermediate nodes having valid routes to the destination, or the destination node itself, are allowed to send RouteReply packets to the source.

Every intermediate node, while forwarding a RouteRequest, enters the previous node address and its BcastID. A timer is used to delete this entry in case a RouteReply is not received before the timer expires. This helps in storing an active path at the intermediate node as AODV does not employ source routing of data packets. When a node receives a RouteReply packet, information about the previous node from which the packet was received is also stored in order to forward the data packet to this next node as the next hop toward the destination.

## Advantages

The main advantage of this protocol is that routes are established on demand and destination sequence numbers are used to find the latest route to the destination. The connection setup delay is lower. It creates no extra traffic for communication along existing links. Also, distance vector routing is simple, and doesn't require much memory or calculation.

## Disadvantages

AODV requires more time to establish a connection, and the initial communication to establish a route is heavier than some other approaches. Also, intermediate nodes can lead to inconsistent routes if the source sequence number is very old and the intermediate nodes have a higher but not the latest destination sequence number, thereby having stale entries. Also multiple RouteReply packets in response to a single RouteRequest packet can lead to heavy control overhead. Another disadvantage of AODV is that the periodic beaconing leads to unnecessary bandwidth consumption.

## Dynamic Source Routing

Dynamic Source Routing (DSR) is a routing protocol for wireless mesh networks. It is similar to AODV in that it forms a route on-demand when a transmitting computer requests one. However, it uses source routing instead of relying on the routing table at each intermediate device. Many successive refinements have been made to DSR, including DSRFLOW.

Determining source routes requires accumulating the address of each device between the source and destination during route discovery. The accumulated path information is cached by nodes processing the route discovery packets. The learned paths are used to route packets. To accomplish source routing, the routed packets contain the address of each device the packet will traverse. This may result in high overhead for long paths or large addresses, like IPv6 (Internet Protocol version 6). To avoid using source routing, DSR optionally defines a flow id option that allows packets to be forwarded on a hop-by-hop basis.

This protocol is truly based on source routing whereby all the routing information is maintained (continually updated) at mobile nodes. It has only two major phases which are Route Discovery and Route Maintenance. Route Reply would only be generated if the message has reached the intended destination node (route record which is initially contained in Route Request would be inserted into the Route Reply).

To return the Route Reply, the destination node must have a route to the source node. If the route is in the Destination Node's route cache, the route would be used. Otherwise, the node will reverse the route based on the route record in the Route Reply message header (this requires that all links are symmetric). In the event of fatal transmission, the Route Maintenance Phase is initiated whereby the Route Error packets are generated at a node. The erroneous hop will be removed from the node's route cache, all routes containing the hop are truncated at that point. Again, the Route Discovery Phase is initiated to determine the most viable route.

Dynamic source routing protocol (DSR) is an on-demand protocol designed to restrict the bandwidth consumed by control packets in ad hoc wireless networks by eliminating the periodic table-update messages required in the table-driven approach. The major difference between this and the other on-demand routing protocols is that it is beacon-less and hence does not require periodic hello packet (beacon) transmissions, which are used by a node to inform its neighbors of its presence. The basic approach of this protocol (and all other on-demand routing protocols) during the route construction phase is to establish a route by flooding RouteRequest packets in the network. The destination node, on receiving a RouteRequest packet, responds by sending a RouteReply packet back to the source, which carries the route traversed by the RouteRequest packet received. Consider a source node that does not have a route to the destination. When it has data packets to be sent to that destination, it initiates a RouteRequest packet. This RouteRequest is flooded throughout the network. Each node, upon receiving a RouteRequest packet, rebroadcasts the packet to its neighbors if it has not forwarded it already, provided that the node is not the destination node and that the packet's time to live (TTL) counter has not been exceeded. Each RouteRequest carries a sequence number generated by the source node and the path it has traversed.

A node, upon receiving a RouteRequest packet, checks the sequence number on the packet before forwarding it. The packet is forwarded only if it is not a duplicate RouteRequest. The sequence number on the packet is used to prevent loop formations and to avoid multiple transmissions of the same RouteRequest by an intermediate node that receives it through multiple paths. Thus, all nodes except the destination forward a RouteRequest packet during the route construction phase. A destination node, after receiving the first RouteRequest packet, replies to the source node through the reverse path the RouteRequest packet had traversed.

Nodes can also learn about the neighboring routes traversed by data packets if operated in the promiscuous mode (the mode of operation in which a node can receive the packets that are neither broadcast nor addressed to itself). This route cache is also used during the route construction phase. If an intermediate node receiving a RouteRequest with a route to the destination node in its route cache, then it replies to the source node by sending a RouteReply with the entire route information from the source node to the destination node.

## Advantages

This protocol uses a reactive approach which eliminates the need to periodically flood the network with table update messages which are required in a table-driven approach. In a reactive (on-demand) approach such as this, a route is established only when it is required and hence the need to find routes to all other nodes in the network as required by the table- driven approach is eliminated. The intermediate nodes also utilize the route cache information efficiently to reduce the control overhead.

## Disadvantages

The disadvantage of this protocol is that the route maintenance mechanism does not locally repair a broken link. Stale route cache information could also result in inconsistencies during the route reconstruction phase. The connection setup delay is higher than in table- driven protocols. Even though the protocol performs well in static and low-mobility environments, the performance degrades rapidly with increasing mobility. Also, considerable routing overhead is involved due to the source-routing mechanism employed in DSR. This routing overhead is directly proportional to the path length.

## Dynamic MANET On-demand Routing

The DYMO routing protocol is a successor to the popular Ad hoc On-Demand Distance Vector (AODV) routing protocol and shares many of its benefits. It is, however, slightly easier to implement and designed with future enhancements in mind. DYMO can work as both a proactive and as a reactive routing protocol, i.e. routes can be discovered just when they are needed. In any way, to discover new routes the following two steps take place:

- A special "Route Request" (RREQ) messages is broadcast through the MANET. Each RREQ keeps an ordered list of all nodes it passed through, so every host receiving an RREQ message can immediately record a route back to the origin of this message.

- When an RREQ message arrives at its destination, a "Routing Reply" (RREP) message will immediately get passed back to the origin, indicating that a route to the destination was found. On its way back to the source, an RREP message can simply backtrace the way the RREQ message took and simultaneously allow all hosts it passes to record a complementary route back to where it came from.

So as soon as the RREP message reaches its destination, a two-way route was successfully recorded by all intermediate hosts, and exchange of data packets can commence.

Example:

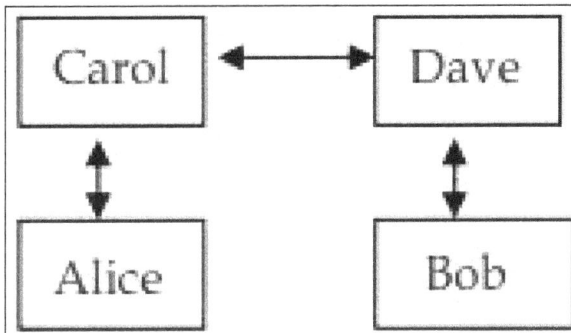

Step 1:

- Alice wants to exchange data with Bob.

- Alice does not know a route to Bob yet, so it broadcasts a new RREQ for a route to Bob containing only information about itself.

Step 2:

- Carol receives Alice's RREQ, remembers the contained information about how to reach Alice (directly), then appends information about itself and re-broadcasts the packets.

Step 3:

- Dave receives Carol's RREQ, remembers the contained information about how to reach Carol (directly) and Alice (via Carol), then appends information about itself and rebroadcasts the packet.

- At the same time, Alice also receives Carol's RREQ. Closer examination of the contained information reveals that even the very first information block how to reach itself, Alice is of no use. It thus discards the RREQ and does not re-broadcast it as Dave did.

Step 4:

- Bob receives Dave's RREQ and remembers the contained information about how to reach Dave (directly), Carol (via Dave) and Alice (also via Dave). Realizing that he is the target of the RREQ he creates an RREP containing information about itself. He marks the RREP bound for Alice and knowing that Dave can somehow reach Alice sends it to Dave.

- Again, at the same time, Carol also receives Dave's RREQ, but following the same logic as Alice before ignores it.

Step 5:

- Dave receives the RREP to Alice sent by Bob, remembers the information on how to reach Bob (directly), appends information about itself and knowing that Alice can be reached via Carol, sends it to Carol.

Step 6:

- Carol receives the RREP to Alice sent by Dave, remembers the contained information on how to reach Dave (directly) and Bob (via Dave), then appends information about itself and knowing that Alice can be reached directly, sends it to Alice.

Step 7:

- Alice receives the RREP sent to her by Carol and remembers all information on how to reach Carol (directly), Dave (via Carol) and most importantly Bob (also via Carol). Now knowing how to reach Bob she can finally send her data packet for him to Carol.

Step 8:

- Carol receives the data packet for Bob from Alice. Because she knows Dave can reach Bob she forwards the packet to him.

Step 9:

- Dave receives the data packet for Bob. Because he knows Bob can be directly reached by him, he forwards the packet to him.

Step 10:

- Bob receives the data packet. Still knowing how to reach Alice, he could now respond with one of his own, and the process repeats until communications are complete or the network changes (e.g. Carol leaves or Eileen joins), where it may be necessary to search the network again for a route.

## On-demand Cache Routing Protocol

This protocol presents an efficient algorithm for route discovery, route management and mobility handling for on-demand routing. It is called as "on-demand cache routing" (ODCR) algorithm since it applies caches in each node to improve the routing performance.

In the MANET, each node equips L-1 (level 1 or primary) and L-2 (level 2 or secondary) caches. Usually, the size of L-1 cache is about 64 to 256 KB and L-2 cache is about 256 KB to 2MB). For memory address mapping, they use 2-, 4- or 8-way set associative scheme. Each data entry in a cache is called a "cache line". Most caches use the least-recently-used (LRU) policy for cache line replacement. All cache lines can be searched in parallel in a few processor cycles. This is an important reason why many routing protocols adopted cache for route management. This cache is called as "route cache" because it stores the routing information in the network.

For the initial settings of a MANET, this protocol assumes (1) the communication media among nodes (e.g. laptop computers) is RF; (2) each node has an identification (ID) number; (3) each node keeps an ID list in its own cache; (4) the wireless links in the network are symmetric (i.e. bi-directional transmission); and (5) the network is scalable and heterogeneous. This means the number of nodes in the network is changeable anytime, and the processor architecture, transmission radius and battery life of each node can be different.

Here, we only present the main algorithm (ODCR). For detail operations of sub algorithms RDA and MHA mentioned in Algorithm ODCR below.

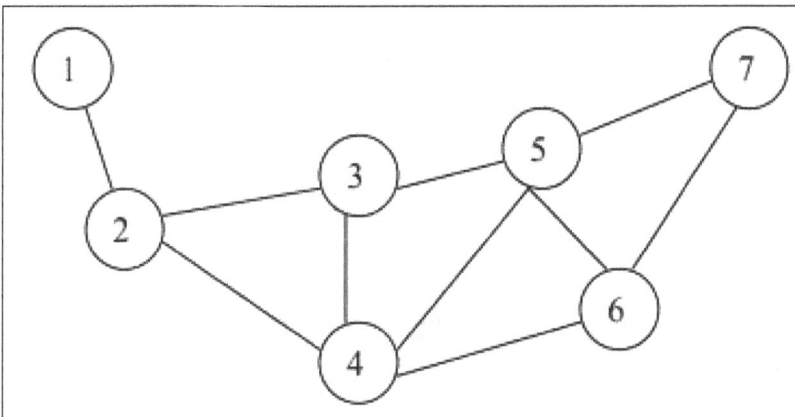

A simple MANET, where 1, 2, 3, 4, 5, 6 and 7 are node IDs and solid edges arewireless links within the RF transmission radius of each node. For example, node 5 can transmits packets to nodes 3, 4, 6 and 7. In this MANET, each node has an ID list (1, 2, 3, 4, 5, 6, 7).

Algorithm: On-Demand Cache Routing (ODCR).

Inputs: Node identifications (IDs) in the MANET.

Outputs: Transmitted data packets on the network.

**Begin**

- If any node in the network wants to send a data packet, at first it has to search the best route (usually the least hop-count route) from its cache. If the route does not exist, go to Step 2. Otherwise (i.e. the route exists) go to Step 3.

- The source node looks up the destination node in its ID list. Then it executes the Route Discovery Algorithm (RDA) to create the best route to its destination node in the network. For instance, the best route from node 1 to node 6 is {1,2,4,6}.

- The source node attaches its ID, destination node ID and the packet number to each data packet, and sends the packet to the destination node along the best route.

- Each intermediate node uses the best route to the destination node in its cache to forward the data packet to the next or destination node.

- If any node leaves from, joins to, or moves around the network, it has to execute the Mobility Handling Algorithm (MHA) to notify other nodes about this change and to update their own route information in their caches.

- Repeat Steps 1 to 5 until the whole network is terminated.

End of On-Demand Cache Routing.

In conclusion, this protocol proposed an efficient on-demand routing algorithm, called ODCR, for route discovery and management, and mobility handling. The ODCR algorithm applied the content-addressable and LRU replacement features in L-1 and L-2 caches for route table creation, updating, and maintenance. The ODCR algorithm with duel-level route caches solved most problems in on-demand routing, such as route tables in "slow" main memory, long connection setup delay, broken link repairing, huge routing overhead for long routes, lengthy data packet in source routing, sending beacons ("hello packets") periodically, control overhead for complicated IDs in data packets, to setup TTL (time-tolive) in a packet or a route path, and to update the stale routes in the route table or cachefrequently.

The simulation results showed that the ODCR algorithm outperforms AODV, DSR (Dynamic Source Routing) and CSOR (Cache Scheme in On-Demand Routing) in packet delivery rate, average end-to-end delay and average routing load.

## Hybrid Routing Protocols

This type of protocols combines the advantages of proactive and reactive routings. The routing is initially established with some proactively prospected routes and then serves the demand from additionally activated nodes through reactive flooding. The choice for one or the other method requires predetermination for typical cases. The main disadvantages of such algorithms are:

- Advantage depends on amount of nodes activated.

- Reaction to traffic demand depends on gradient of traffic volume.

## Zone Routing Protocol

Zone Routing Protocol (ZRP) was the first hybrid routing protocol with both a proactive and a reactive routing component. ZRP was first introduced by Haas in 1997. ZRP is proposed to reduce the control overhead of proactive routing protocols and decrease the latency caused by routing discover in reactive routing protocols. ZRP defines a zone around each node consisting of its k-neighborhood (e.g. k=3). That is, in ZRP, all nodes within k-hop distance from node belong to the routing zone of node.

ZRP is formed by two sub-protocols, a proactive routing protocol: Intra-zone Routing Protocol (IARP), is used inside routing zones and a reactive routing protocol: Inter-zone Routing Protocol (IERP), is used between routing zones, respectively. A route to a destination within the local zone can be established from the proactively cached routing table of the source by IARP. Therefore, if the source and destination is in the same zone, the packet can be delivered immediately. Most of the existing proactive routing algorithms can be used as the IARP for ZRP.

For routes beyond the local zone, route discovery happens reactively. The source node sends a route requests to its border nodes, containing its own address, the destination address and a unique sequence number. Border nodes are nodes which are exactly the maximum number of hops to the defined local zone away from the source. The border nodes check their local zone for the destination. If the requested node is not a member of this local zone, the node adds its own address to the route request packet and forwards the packet to its border nodes. If the destination is a member of the local zone of the node, it sends a route reply on the reverse path back to the source. The source node uses the path saved in the route reply packet to send data packets to the destination.

## Order One Network Protocol

The Order One MANET Routing Protocol (OORP) is an algorithm for computer communicating by digital radio in a mesh network to find each other, and send messages to each other along a reasonably efficient path. It was designed for, and promoted as working with wireless mesh networks. OORP can handle hundreds of nodes, where most other protocols handle less than a hundred. OORP uses hierarchical algorithms to minimize the total amount of transmissions needed for routing. Routing overhead is only about 1% to 5% of node to node bandwidth in any network and does not grow as the network size grows.

The basic idea is that a network organizes itself into a tree. Nodes meet at the root of the tree to establish an initial route. The route then moves away from the root by cutting corners, as ant-trails do. When there are no more corners to cut, a nearly optimum route exists. This route is continuously maintained. Each process can be performed with localized minimal communication, and very small router tables. OORP requires about 200K of memory. A simulated network with 500 nodes transmitting at 200 bytes/second organized itself in about 20 seconds. As of 2004, OORP was patented or had other significant intellectual property restrictions.

Assumptions:

Each computer or "node" of the network has a unique name. At least one network link and a computer with some capacity hold a list of neighbors.

## Organizing a Tree

The network nodes form a hierarchy by having each node select a parent. The parent is a neighbor node that is the next best step to the most other nodes. This method creates a hierarchy around nodes that are more likely to be present, and which have more capacity, and which are closer to the topological center of the network. The memory limitations of a small node are reflected in its small routing table, which automatically prevents it from being a preferred central node. At the top, one or two nodes are unable to find nodes betterconnected than themselves, and therefore become parents of the entire network. The hierarchy-formation algorithm does not need a complex routing algorithm or large amounts of communication.

## Routing

All nodes push a route to themselves to the root of the tree. A node wanting a connection can therefore push a request to the root of the tree, and always find a route. The commercial protocol uses Dijkstra's algorithm to continuously optimize and maintain the route. As the network moves and changes, the path is continually adjusted.

## Advantages

Assuming that some nodes in the network have enough memory to know of all nodes in the network, there is no practical limitation to network size. Since the control bandwidth is defined to be less than 5% regardless of network size, the amount of control bandwidth required is not supposed to increase as network size grows. The system can use nodes with small amounts of memory.

The network has a reliable, low-overhead way to establish that a node is not in the network. This is a valuable property in ad-hoc mesh networks. Most routing protocols scale either by reducing proactive link-state routing information or reactively driving routing by connection requests. OORP mixes the proactive and reactive methods. Properly configured, an OORP net can theoretically scale to 100,000's of nodes and can often achieve reasonable performance even though it limits routing bandwidth to 5%.

## Disadvantages

Central nodes have an extra burden because they need to have enough memory to store information about all nodes in the network. At some number of nodes, the network will therefore cease to scale. If all the nodes in the network are low capacity nodes the network may be overwhelmed with change. This may limit the maximum scale. However, in real world networks, the farther away from the edge nodes the more the bandwidth grows. These critiques may have no practical effect. For example, consider a low bandwidth 9.6Kbit/second radio. If the protocol was configured to send one packet of 180 bytes every 5 seconds, it would consume 3% of overall network bandwidth. This one packet can contain up to 80 route updates. Thus even in very low bandwidth network the protocol is still able to spread a lot of route information. Given a 10Mbit connection, 3% of the bandwidth is 4,100 to 16,000 route updates per second. Since the protocol only sends route updates for changes, it is rarely overwhelmed.

The other disadvantage is that public proposals for OORP do not include security or authentication. Security and authentication may provided by the integrator of the protocol. Typical security measures include encryption or signing the protocol packets and incrementing counters to prevent replay attacks.

## Global On-demand Routing protocol

The Global On-Demand Routing (GOR) is a clever hybrid routing protocol for the MANET. To simplify simulations in GOR, it assumes (1) all nodes are homogeneous; (2) the transmission range of each node is k; and (3) each node has an ID and a pair of positive x and y coordinates to represent its location in the network. The main algorithm for the GOR protocol is described below. For detail operations of sub-algorithms DFA and NRA in GOR protocol.

Algorithm: GOR Protocol:

Inputs: The ID and (x, y) coordinates of each node.

Outputs: Destination nodes receive data packets from sources nodes.

## Begin

- Select a center or near-center node in the initial network as the root node (RN).

- The RN runs the Double-Flooding Algorithm (DFA) to create the location table (LT), sorts the LT by IDs in ascending order, and broadcasts the LT to each node in the network.

- Each node uses the LT to generate its own distance table (DT) concurrently. Then, each node marks any distance that is longer than the transmission range k in the DT as "∞" (infinity).

- Each node calls the Dijkstra's Algorithm to generate the one-to-all shortest-path table (SPT) concurrently.

- If a new node joined to the network, an existing node moved out of the transmission range of its any neighbor nodes, or an existing node left from the network, then it calls the Node-Reorganization Algorithm (NRA) to ask other nodes to update (or mark as "new" nodes if any) their own LT for these changes consequently.

- If any node wants to send packets via or to the above joined or moved nodes, it has to (1) use the updated LT in Step 5 to update its DT (or mark the "∞" distances if any); (2) run the Dijkstra's algorithm again to update its SPT; (3) reset all nodes in the LT to "old" nodes; and (4) follows the paths in the new SPT to send packets to its destination nodes.

- If network topology changed again, repeat steps 5 and 6 until the whole network dismissed.

End of GOR Protocol.

Figure below shows some shortest paths within the transmission range k for node 1. In this figure, the shortest path between nodes 1 and 6 is (1, 3, 6) not (1, 6) because node 6 locates outside the circular transmission range k of node 1. Note we have marked all "∞" distances in steps 3 and 6 respectively in the main algorithm (Algorithm GOR Protocol).

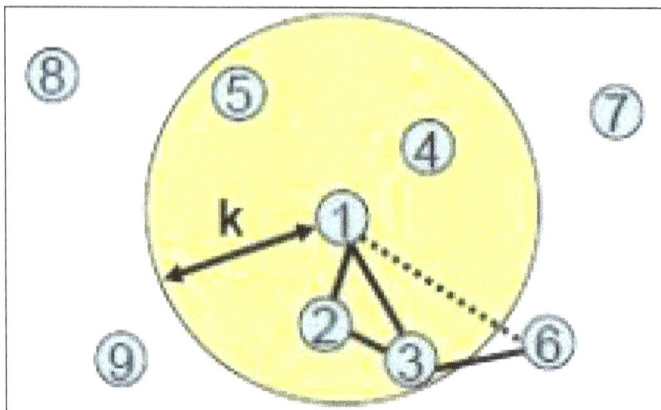

Sample shortest paths in a MANET.

This algorithm proposed a hybrid global on-demand routing (called GOR) protocol for mobile ad hoc networks. This protocol does not update the routing tables immediately if any node changed its status in the network, such as movement, addition or deletion.

Instead, it only handles a node whose move changed the MANET topology or whose move distance is greater than the transmission range k. This critical strategy prevents other nodes from updating the routing tables frequently and hence reducing unnecessary computation and node-reorganization overheads dramatically.

The GOR protocol not only keeps the advantages of proactive and reactive protocols, but also improves the sub-optimal routing overhead and memory consuming problems in local hybrid protocols. Because this protocol retains high packet delivery rate and low end-to-end delay as the DSDV and WRP protocols, and low routing load as the AODV and DSR protocols.

## MANET Routing Protocols for IPv6

It is possible that all the IP version 4 (IPv4) addresses will be allocated in next decade. The transition from IP version 4 to IP version 6 (IPv6) will become an important issue in computer networks and Internet in recent years. Therefore, here, we introduce IPv6, mobile IPv6, and two popular MANET routing protocols, OLSR and AODV, for IPv6 networks.

## Introduction to IPv6 and Mobile IPv6

Internet is built upon a protocol suite called TCP/IP. This abbreviation stands for Transmission Control Protocol, and Internet Protocol. When your computer communicates with the Internet a unique IP address is used to transfer and receive information. Yesterdays IP standard is called IPv4. Each IPv4 address contains 32 binary bits. That is the total address in IPv4 is $2^{32}$ only. Sadly most ISPs and services still only deliver this ancient technology standardized in September 1981. So far, most of IPv4 addresses are already tied up and the Internet is simply running out of IPs. The address shortage problem is aggravated by the fact that portions of the IP address space have not been efficiently allocated.

IPv6 (Internet Protocol version 6) gives citizens the opportunity to become real Internet participants. IPv4 makes citizens into passive consumers who are only able to connect to compartmentalized networks run by companies or governments. This is why the establishment does not want IPv6. Each IPv6 address contains 128 binary bits. This means there are $2^{128}$ unique addresses in IPv6. This huge amount of IP addresses may be able to serve the Internet till the end of this century.

Mobile IPv6 is the implementation of the IP mobility (Mobile IP) methods and protocols on an Internet Protocol version 6 (IPv6) network. Like its IPv4 counterpart, it is designed to permit IP devices to roam between different networks without losing IP connectivity by maintaining a permanent Internet Protocol (IP) address.

The key benefit of Mobile IPv6 is that even though the mobile node changes locations and addresses, the existing connections through which the mobile node is communicating are maintained. To accomplish this, connections to mobile nodes are made with a specific address that is always assigned to the mobile node, and through which the mobile node is always reachable. Mobile IPv6 provides Transport layer connection survivability when a node moves from one link to another by performing address maintenance for mobile nodes at the Internet layer.

## OLSR for IPv6 Networks

We summarize the proposed issues and necessary changes to adapt OLSR to IPv6:

- Addressing: IPv6 introduce several changes, some more conceptual than others. Changes include the diffusion of data packets and existing multiple addresses of Interfaces.

- Protocol changes: The OLSR specification gives the protocol format message for IPv4 packets, but some additional changes are proposed.

- Neighbor discovery: It is described how the neighbor discovery mechanism of IPv6 still operates properly.

- Autoconfiguration: It is loosely related to addressing, the ability for an IPv6 node to self-configure its addressed yields numerous challenges and had been the subject of elaborate research as seen previously.

## IPv6 Ad-Hoc Addressing Issues

Several changes are required due to various novelties introduced by IPv6 itself.

- Interface Addresses: The chosen solution here is to consider a MANET as a single site-local network, and to use site-local prefix with a fixed 16 bits subnet called OLSR_SUBNET. Then, an OLSR node will perform link-local address autoconfiguration, and upon success, will automatically configure for each of its OLSR interfaces. The sitelocal address with that subnet (FEC0:0:0:OLSR_SUBNET::/64) will run the OLSR protocol using it.

- OSLR Diffusion Addresses: In order to reach all the nodes present on the link to get the same effect as in IPv4, this paper proposed that a multicast address ALL_OLSR_NODES is used for the destination address. The ALL_OLSR_NODES could be taken as ALL_LINK_NODES (FF01::1). Also since a node has several interface addresses, the paper proposed that the site-local addresses are used as source addresses.

## Diffusing Non-OLSR Packets

Since MANETs are multi-hop routing networks, in order to flood packets to all nodes, retransmissions are usually needed. With OLSR, packets are retransmitted hop by hop to the direct neighborhood by using MPRs (multipoint relays). In the other hand, for any applications, a direct multicast on the local "link" is performed and such packets are never routed. For instance, it is also in the case for most of IPv6 messages for neighbor discovery and autoconfiguration. This relies on the assumption that being on the same network is equivalent to being on same link, an assumption which doesn't hold in MANET networks.

As a result, in a multi-hop network, by default, this kind of messages will not be delivered to all nodes. Here, we proposed two solutions to diffuse non-OLSR packets to all nodes:

- Encapsulate the packets in specific OLSR messages, and use the MPR flooding.

- Use of a new multicast address called ALL-MANET_NODES, instead of the ALL_LINK_NODES.

## Changes to the OLSR Routing Protocol

- OLSR Packet format: The essential change needed for the existing OLSR packet format is to replace the IPv4 addresses with the IPv6 addresses in all messages, as highlighted in the OLSR specification.

- Multiple Interface Addresses: In IPv6, an interface can have several addresses. Here, we proposed an OLSR node, for each interface, will have:

  - A link-local address: This address is usually obtained by autoconfiguration. It is temporary used as the source address for OLSR packets before autoconfiguration is completed.

  - A site-local address: This is derived from the link-local address, in the fixed subnet OLSR_SUBNET for site-local prefix. This address is permanently used as the source for all OLSR packets, once autoconfiguration is completed.

  - Any number (possibly zero) of additional global or site local unicast addresses, which are automatically or manually configured.

## Neighbor Discovery

In IPv6, nodes (hosts and routers) use Neighbor Discovery [Narten1998] to determine the MAC addresses for neighbors on attached links and to quickly purge invalid cache values. Hosts also use Neighbor Discovery to find neighboring routers that are willing to forward packets on their behalf. Finally, nodes use the protocol to actively keep track of which neighbors are reachable and which are not, and to detect changed MAC addresses.

Routing table in the OLSR indicates the next hop for each reachable destination in the network. This next hop is one of the direct neighbors. This means that the neighbor solicitation for address resolution will work without any modification. In OLSR, gateways declare themselves to the entire network periodically. The neighbor discovery is adapted to OLSR. Consequently it is not necessary to do any modification to the classical procedure.

## Autoconfiguration

IPv6 Stateless Address Autoconfiguration is based on several steps: after the creation of a link local address, the node must check whether the address is already in use by another interface of another node, somewhere in the network. In wired network, this means that all the links of the attached interfaces of the node are probed. If the address is not unique the process is interrupted, otherwise the autoconfiguration was successful and the address may be safely used.

In a MANET, the nodes on the links of the attached interfaces would include only the nodes with an interface within radio reach of the transmitter and not all the participating nodes.

Hence, the uniqueness of the address is not guaranteed if the classical DAD (Duplicate Address Detection) procedure is applied. Here, we proposed an algorithm, following the philosophy of the IPv6 DAD, to perform autoconfiguration in an OLSR network. The algorithm includes reactive probing (i.e. sending a request to the whole network and waiting for a possible answer), proactive checking (i.e. checking periodically for duplicate addresses) and collision resolution (i.e. what should be done upon detection of duplicate addresses).

## Ad hoc On-demand Distance Vector Routing for IPv6

The operation of AODV for IPv6 is intended to mirror the operation of AODV for IPv4, with changes necessary to allow for transmission of 128-bit addresses in IPv6 instead of the traditional 32-bit addresses in IPv4.

## Route Request Message Format

The format of the IPv6 Route Request message (RREQ) contains the same fields with the same functions as the RREQ message defined for IP version 4, except as follows:

- Destination IP Address: The 128-bit IPv6 address of destination for which a route is desired.

- Source IP Address: The 128-bit IPv6 address of the node which originated the Route Request.

Note, the order of the fields has been changed to enable alignment along the 128-bit boundaries.

## Route Reply Message Format

The format of the IPv6 Route Reply message (RREP) contains the same fields with the same functions as the RREP message defined for IP version 4, except as follows:

- Prefix Size: The Prefix Size is 7 bits instead of 5, to account for the 128-bit IPv6 address space.

- Destination Sequence Number: The destination sequence number associated to the route.

- Destination IP Address: The 128-bit IP address of the destination for which a route is supplied.

- Source IP Address: The 128-bit IP address of the source node which issued the RREQ for which the route is supplied.

Note, the order of the fields has been changed for better alignment.

## Route Error Message Format

The format of the Route Error (RERR) message is identical to the format for the IPv4 RERR message except that the IP addresses are 128 bits, not 32 bits.

## Route Reply Acknowledgment Message Format

The RREP-ACK message is used to acknowledge receipt of an RREP message. It is used in cases where the link over which the RREP message is sent may be unreliable. It is identical in format to the RREP-ACK message for IPv4.

## AODV for IPv6 Operation

The handling of AODV for IPv6 messages analogous to the operation of AODV for IPv4, except that the RREQ, RREP, RERR, and RREP-ACK messages described above are to be used instead; these messages have the formats appropriate for use with 128-bit IPv6 addresses.

# Issues and Security | 5

Mobile communication network security is an essential component in mobile computing as it provides security of personal and business information stored in smartphones. Some of its aspects are Bluetooth security, wired equivalent privacy, Wi-Fi protected access, 4G and LTE network security. This chapter discusses the related aspects of security and issues of mobile computing in detail.

## Issues in Mobile Computing

Mobile computing has its fair share of security concerns as any other technology. Due to its nomadic nature, it's not easy to monitor the proper usage. Users might have different intentions on how to utilize this privilege. Improper and unethical practices such as hacking, industrial espionage, pirating, online fraud and malicious destruction are some but few of the problems experienced by mobile computing.

Another big problem plaguing mobile computing is credential verification. As other users share username and passwords, it poses as a major threat to security. This being a very sensitive issue, most companies are very reluctant to implement mobile computing to the dangers of misrepresentation.

The problem of identity theft is very difficult to contain or eradicate. Issues with unauthorized access to data and information by hackers, is also an enormous problem.

Outsiders gain access to steal vital data from companies, which is a major hindrance in rolling out mobile computing services.

No company wants to lay open their secrets to hackers and other intruders, who will in turn sell the valuable information to their competitors. It's also important to take the necessary precautions to minimize these threats from taking place. Some of those measures include:

- Hiring qualified personnel.

- Installing security hardware and software.

- Educating the users on proper mobile computing ethics.

- Auditing and developing sound, effective policies to govern mobile computing.

- Enforcing proper access rights and permissions.

These are just but a few ways to help deter possible threats to any company planning to offer mobile computing. Since information is vital, all possible measures should be evaluated and implemented for safeguard purposes.

In the absence of such measures, it's possible for exploits and other unknown threats to infiltrate and cause irrefutable harm. These may be in terms of reputation or financial penalties. In such cases, it's very easy to be misused in different unethical practices.

If these factors aren't properly worked on, it might be an avenue for constant threat. Various threats still exist in implementing this kind of technology.

## Bluetooth Security

In Bluetooth, there are three security modes:

- Security Mode 1: In this mode, the device does not implement any security procedures, and allows any other device to initiate connections with it.

- Security Mode 2: In mode 2, security is enforced after the link is established, allowing higher level applications to run more flexible security policies.

- Security Mode 3: In mode 3, security controls such as authentication and encryption are implemented at the Baseband level before the connection is established. In this mode, Bluetooth allows different security levels to be defined for devices and services.

### Devices

Devices are other Bluetooth devices that wish to use the services you provide. These devices are divided into two categories: Trusted and untrusted.

- Trusted devices have a fixed relationship with your device. They generally have unrestricted access to all services, although this can be refined to set different access policies for each service.

- Untrusted devices have no permanent fixed relationship with your device (but may have a temporary one) or have a permanent relationship but are considered untrustworthy. They are generally limited as to the services they can access.

## Services

Services are the services you provide to other devices, and are divided into three categories:

- Services that require authorisation and authentication: Only trusted devices have automatic access; all other devices require manual authorization.

- Services that require authentication only: authorisation is not required.

- Services that are open to all: access to services is granted without the need for approval.

## Service Level Security

In security mode 2, security is handled by higher level applications rather than at the link level, and is enforced after the communication is established.

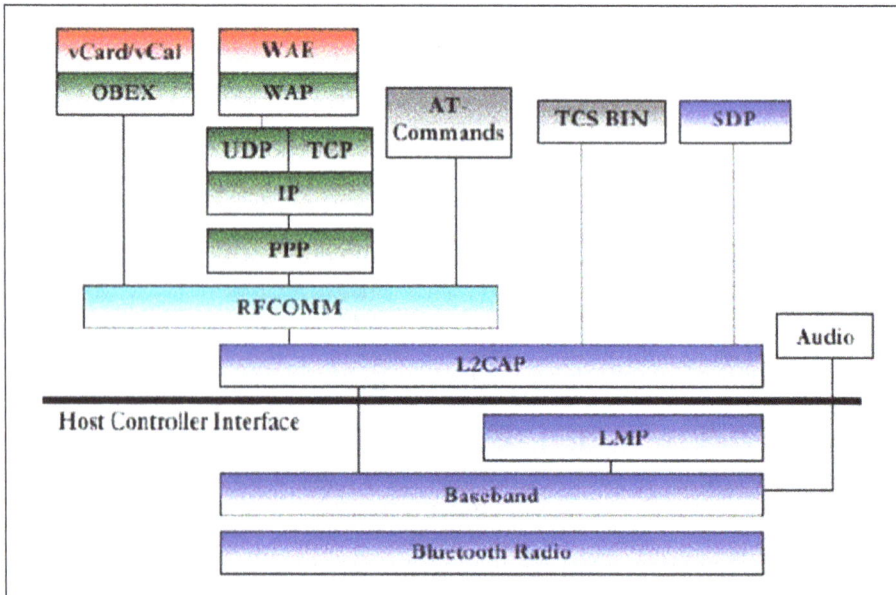

The Bluetooth architecture.

Because Bluetooth uses the RFCOMM protocol it is able to use existing protocols such as TCP, UDP and WAP, and can use the security measures built into these.

## Link Level

In security mode 3, security is enforced before a communications link is established. Security in Bluetooth, like in other networks is based on authentication and encryption.

In Bluetooth security there are four main identifiers:

- 48 bit unique IEEE Bluetooth device address (BD_ADDR).

- 128 bit Link key, used for authentication.

- 8-128 bit symmetric encryption key.

- 128 bit random numbers (RAND) generated as required.

## Authentication

To authenticate devices in Bluetooth, a link key is generated for the connection, followed by a challenge-response strategy to ensure that the claimant device knows the link key. There are four types of link key defined in the Bluetooth specification: unit keys, initialization keys, combination keys and master keys. In addition, Bluetooth defines four algorithms (E1, E21, E22 and E3) for key generation and authentication. They are all based on the SAFER+ block cipher algorithm.

## Unit Key

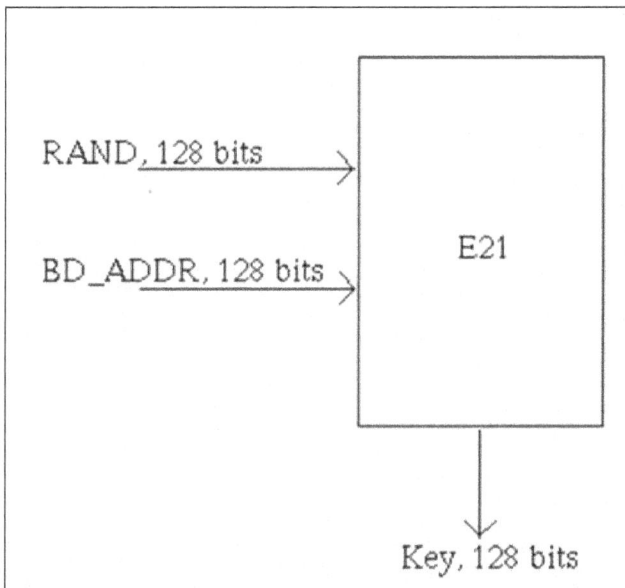

The E21 key generation algorithm used in generating unit and combination keys.

The unit key is generated in the device when it is first turned on, using the E21 key generating algorithm. Once generated, it is stored in the non-volatile memory of the device and rarely changes. The unit key can be used as a link key between two devices, however it is a last resort since sharing the unit key allows the device to be 'spoofed'. For example, if device A communicates with device B using A's unit key as the link key, device B then has a copy of A's unit key. Device B can then pretend to be device A and start a communication with device C, or can intercept and decode communications between A and C if A's unit key is used as the link key. In general, the unit key is only used as a link key when one of the devices has very limited memory and is not capable of remembering any extra keys.

## Initialization Key

The initialization key is needed for two devices who have never communicated before. A PIN code of between 1 and 16 octets in length is entered into both devices, and an initialization key is generated using the E22 key generating algorithm. The algorithm is given the PIN (augmented with the BD_ADDR of the claimant device), and a 128bit random number from the verifying device. This key is then used for key exchange during the generation of a link key after which it is discarded.

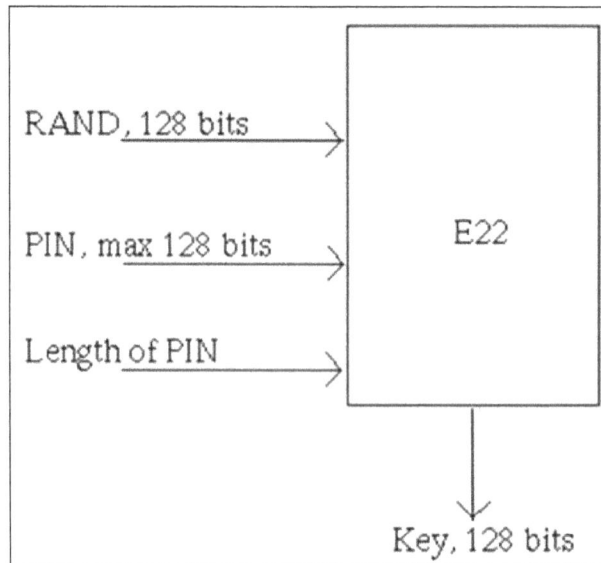

The E22 key generation algorithm, used in generating initialization and master keys.

## Combination Key

The combination key is generated using another key, usually the initialization key, and is the most secure of the four keys. Firstly, each unit generates a temporary key using the same method as used for unit key creation (e.g. device 1 inputs RAND1 and BD_ADDR1 into the E21 algorithm to produce KEY1, device 2 inputs RAND2 and BD_ADDR2 to produce KEY2). Device 1 then XORs its temporary key (KEY1) with the current link key and sends it to device 2. Similarly, device 2 XORs KEY2 and sends it to device 1. Each device then decrypts the random number, and using the BD_ADDR of the other device, calculates the other temporary key. Both devices then XOR KEY1 and KEY2 to produce the combination key.

## Master Key

This key is the only temporary key and is generated when one device needs to communicate with several others (one master, several slaves). Firstly, the master key is created (by the master) by putting two random numbers into the E22 key generation algorithm. The master then sends another random number to each slave. The master and all the slaves then put the current link key and the random number into the key generating algorithm to produce an overlay. The master then XORs the master key with that overlay and transmits it to all of the slaves. The slaves (which each have a copy of the overlay) can then calculate the master key.

Once the link key has been established, Bluetooth uses a challenge-response scheme for authentication, and relies on both parties sharing the same secret key. Firstly, the verifier sends the claimant a 128 bit random number. The claimant inputs this number, the current link key and its own BD_ADDR into the authentication algorithm E1, and sends the output back to the verifier. If the response matches the value calculated by the verifier, the claimant is authenticated. A by-product of the authentication algorithm is the ACO (Authenticated Ciphering Offset) which is stored by both devices for use later on in the encryption process.

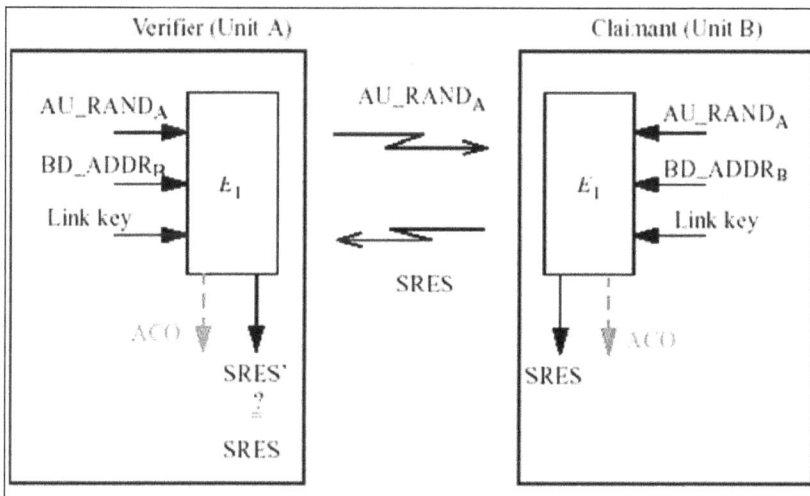

The E1 authentication algorithm.

## Encryption

To encrypt payloads of their packets, Bluetooth devices uses private encryption key derived from the current link key, the 96 bit COF (Ciphering Offset Number based on the ACO generated during the authentication process) and a 128 bit random number.

This encryption key, along with the device clock, the device address and a random number are fed into the E0 stream cipher algorithm (based on the Massey and Rueppel summation stream cipher generator). The encryption key is regenerated every new packet transmission.

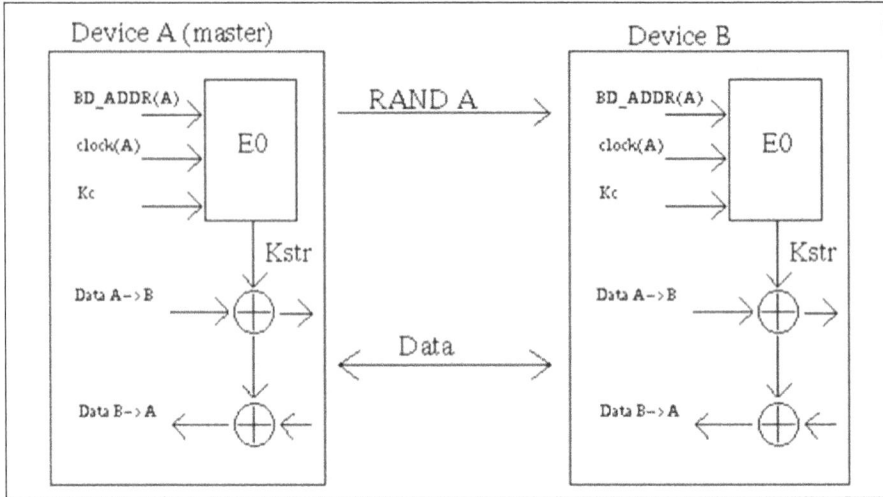

Encryption in Bluetooth.

Since the encryption key size varies from 8 bits to128 bits, the devices have to negotiate the length of the encryption key. Firstly, the master sends a suggested length to the slave. The slave can then accept this, or reject it and send back a suggestion for another length. This continues until a consensus is reached, or until one of the devices aborts the negotiation.

## Weaknesses with Bluetooth

There are three main types of attacks on Bluetooth connections:

## Attacks on the Confidentiality

It has been suggested that the E0 stream cipher can be broken in $2^{66}$ in some circumstances. However, the fact that Bluetooth devices resynchronize the stream cipher every packet means that malicious users attempting to break the encryption in this way would not have sufficient time to succeed.

If a device's unit key is used as the link key, trusted devices can spoof that device and can initiate communications posing as that device, or can intercept communications between that device and another which use the unit key as a link key.

The fact that Bluetooth uses 4 digit pins for the initialization key is a weakness. With only 10000 possibilities, PINs can be exhaustively searched and the key broken. In addition to this, research has shown that 50% of PINs are 0000, putting the security of PINs further into question. One way to avoid this threat is by only pairing devices in secure locations. Given that the range of most Bluetooth devices is only 10m (less when the signal has to travel through walls) this is not too inconvenient, and generally only prohibits users from pairing in busy locations such as stations or airports.

## Attacks on the Availability

Bluetooth uses the unlicensed and heavily used, 2.45Ghz frequency. This makes it susceptible to interference from other devices, such as microwaves. However, to combat this it uses the technique of frequency hopping. This improves the clarity and also discourages casual eavesdropping, since only synchronized devices can communicate.

Another form of denial of service attack, the battery exhaustion attack, is one that Bluetooth devices are susceptible to. Unfortunately, it is very difficult to prevent this kind of attack without restricting the usability of the device. For example, users can choose to turn off Bluetooth until they need it, but this is irritating for the user. Work is underway by the Bluetooth Special Interest Group to find a solution, but currently there is no effective defense.

## Attacks on User Privacy

Another issue with Bluetooth is that of user privacy. Since the address of the device is freely available, once it is associated with an individual it can be used to carry out profiling, and other more questionable forms of monitoring, violating user privacy.

Bluejacking is where users can send anonymous, unsolicited messages to other Bluetooth users Bluetooth in the form of business cards, exploiting the Bluetooth standard of accepting them. Although currently this is a relatively small and innocuous craze, the same property could be used by unscrupulous marketers to target passing customers (and indeed in some places, this is already happening). The combination of these two attacks on privacy could prove to be extremely annoying and intrusive to end users. For example, companies with your details (including the address of your Bluetooth device) on record could bombard you with personalized adverts and offers whenever you are near their store.

## WEP (Wired Equivalent Privacy)

Wired Equivalent Privacy (WEP) is a security algorithm for IEEE 802.11 wireless networks. Introduced as part of the original 802.11 standard ratified in 1997, its intention was to provide data confidentiality comparable to that of a traditional wired network. WEP, recognizable by its key of 10 or 26 hexadecimal digits (40 or 104 bits), was at one time widely in use and was often the first security choice presented to users by router configuration tools.

In 2003 the Wi-Fi Alliance announced that WEP had been superseded by Wi-Fi

Protected Access (WPA). In 2004, with the ratification of the full 802.11i standard (i.e. WPA2), the IEEE declared that both WEP-40 and WEP-104 have been deprecated.

WEP was the only encryption protocol available to 802.11a and 802.11b devices built before the WPA standard, which was available for 802.11g devices. However, some 802.11b devices were later provided with firmware or software updates to enable WPA, and newer devices had it built in.

### Encryption Details

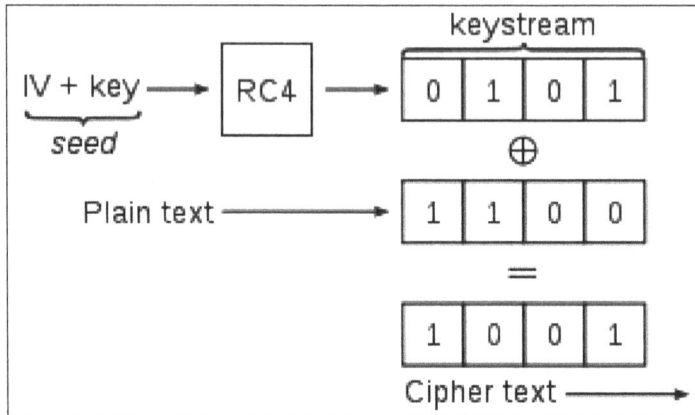

Basic WEP encryption: RC4 key stream XORed with plaintext.

WEP was included as the privacy component of the original IEEE 802.11 standard ratified in 1997. WEP uses the stream cipher RC4 for confidentiality, and the CRC-32 checksum for integrity. It was deprecated in 2004 and is documented in the current standard.

Standard 64-bit WEP uses a 40 bit key (also known as WEP-40), which is concatenated with a 24-bit initialization vector (IV) to form the RC4 key. At the time that the original WEP standard was drafted, the U.S. Government's export restrictions on cryptographic technology limited the key size. Once the restrictions were lifted, manufacturers of access points implemented an extended 128-bit WEP protocol using a 104-bit key size (WEP-104).

A 64-bit WEP key is usually entered as a string of 10 hexadecimal (base 16) characters (0–9 and A–F). Each character represents 4 bits, 10 digits of 4 bits each gives 40 bits; adding the 24-bit IV produces the complete 64-bit WEP key (4 bits × 10 + 24 bits IV = 64 bits of WEP key). Most devices also allow the user to enter the key as 5 ASCII characters (0–9, a–z, A–Z), each of which is turned into 8 bits using the character's byte value in ASCII (8 bits × 5 + 24 bits IV = 64 bits of WEP key); however, this restricts each byte to be a printable ASCII character, which is only a small fraction of possible byte values, greatly reducing the space of possible keys.

A 128-bit WEP key is usually entered as a string of 26 hexadecimal characters. 26 digits of 4 bits each gives 104 bits; adding the 24-bit IV produces the complete 128-bit WEP key (4 bits × 26 + 24 bits IV = 128 bits of WEP key). Most devices also allow the user to enter it as 13 ASCII characters (8 bits × 13 + 24 bits IV = 128 bits of WEP key).

152-bit and 256-bit WEP systems are available from some vendors. As with the other WEP variants, 24 bits of that is for the IV, leaving 128 or 232 bits for actual protection. These 128 or 232 bits are typically entered as 32 or 58 hexadecimal characters (4 bits × 32 + 24 bits IV = 152 bits of WEP key, 4 bits × 58 + 24 bits IV = 256 bits of WEP key). Most devices also allow the user to enter it as 16 or 29 ASCII characters (8 bits × 16 + 24 bits IV = 152 bits of WEP key, 8 bits × 29 + 24 bits IV = 256 bits of WEP key).

## Authentication

Two methods of authentication can be used with WEP: Open System authentication and Shared Key authentication.

In Open System authentication, the WLAN client does not provide its credentials to the Access Point during authentication. Any client can authenticate with the Access Point and then attempt to associate. In effect, no authentication occurs. Subsequently, WEP keys can be used for encrypting data frames. At this point, the client must have the correct keys.

In Shared Key authentication, the WEP key is used for authentication in a four-step challenge-response handshake:

- The client sends an authentication request to the Access Point.

- The Access Point replies with a clear-text challenge.

- The client encrypts the challenge-text using the configured WEP key and sends it back in another authentication request.

- The Access Point decrypts the response. If this matches the challenge text, the Access Point sends back a positive reply.

After the authentication and association, the pre-shared WEP key is also used for encrypting the data frames using RC4.

At first glance, it might seem as though Shared Key authentication is more secure than Open System authentication, since the latter offers no real authentication. However, it is quite the reverse. It is possible to derive the key stream used for the handshake by capturing the challenge frames in Shared Key authentication. Therefore, data can be more easily intercepted and decrypted with Shared Key authentication than with Open System authentication. If privacy is a primary concern, it is more advisable to use Open System authentication for WEP authentication, rather than Shared Key authentication; however, this also means that any WLAN client can connect to the AP. (Both authentication mechanisms are weak; Shared Key WEP is deprecated in favor of WPA/WPA2).

## Weak Security

Because RC4 is a stream cipher, the same traffic key must never be used twice. The purpose of an IV, which is transmitted as plain text, is to prevent any repetition, but a 24-bit IV is not long enough to ensure this on a busy network. The way the IV was used also opened WEP to a related key attack. For a 24-bit IV, there is a 50% probability the same IV will repeat after 5,000 packets.

In August 2001, Scott Fluhrer, Itsik Mantin, and Adi Shamir published a cryptanalysis of WEP that exploits the way the RC4 ciphers and IV are used in WEP, resulting in a passive attack that can recover the RC4 key after eavesdropping on the network. Depending on the amount of network traffic, and thus the number of packets available for inspection, a successful key recovery could take as little as one minute. If an insufficient number of packets are being sent, there are ways for an attacker to send packets on the network and thereby stimulate reply packets which can then be inspected to find the key. The attack was soon implemented, and automated tools have since been released. It is possible to perform the attack with a personal computer, off-the-shelf hardware and freely available software such as aircrack-ng to crack any WEP key in minutes.

Cam-Winget et al. surveyed a variety of shortcomings in WEP. They write "Experiments in the field show that, with proper equipment, it is practical to eavesdrop on WEP-protected networks from distances of a mile or more from the target". They also reported two generic weaknesses:

- The use of WEP was optional, resulting in many installations never even activating it.

- By default, WEP relies on a single shared key among users, which leads to practical problems in handling compromises, which often leads to ignoring compromises.

In 2005, a group from the U.S. Federal Bureau of Investigation gave a demonstration where they cracked a WEP-protected network in three minutes using publicly available tools. Andreas Klein presented another analysis of the RC4 stream cipher. Klein showed that there are more correlations between the RC4 keystream and the key than the ones found by Fluhrer, Mantin and Shamir which can additionally be used to break WEP in WEP-like usage modes.

In 2006, Bittau, Handley, and Lackey showed that the 802.11 protocol itself can be used against WEP to enable earlier attacks that were previously thought impractical. After eavesdropping a single packet, an attacker can rapidly bootstrap to be able to transmit arbitrary data. The eavesdropped packet can then be decrypted one byte at a time (by transmitting about 128 packets per byte to decrypt) to discover the local network IP addresses. Finally, if the 802.11 network is connected to the Internet, the attacker can use 802.11 fragmentations to replay eavesdropped packets while crafting a new IP header onto them. The access point can then be used to decrypt these packets and relay them on to a buddy on the Internet, allowing real-time decryption of WEP traffic within a minute of eavesdropping the first packet.

In 2007, Erik Tews, Andrei Pychkine, and Ralf-Philipp Weinmann were able to extend Klein's 2005 attack and optimize it for usage against WEP. With the new attack it is possible to recover a 104-bit WEP key with probability 50% using only 40,000 captured packets. For 60,000 available data packets, the success probability is about 80% and for 85,000 data packets about 95%. Using active techniques like deauth and ARP re-injection, 40,000 packets can be captured in less than one minute under good conditions. The actual computation takes about 3 seconds and 3 MB of main memory on a Pentium-M 1.7 GHz and can additionally be optimized for devices with slower CPUs. The same attack can be used for 40-bit keys with an even higher success probability.

In 2008 the Payment Card Industry (PCI) Security Standards Council updated the Data Security Standard (DSS) to prohibit use of WEP as part of any credit-card processing after 30 June 2010, and prohibit any new system from being installed that uses WEP after 31 March 2009. The use of WEP contributed to the TJ Maxx parent company network invasion.

## Remedies

Use of encrypted tunneling protocols (e.g. IPSec, Secure Shell) can provide secure data transmission over an insecure network. However, replacements for WEP have been developed with the goal of restoring security to the wireless network itself.

## 802.11i (WPA and WPA2)

The recommended solution to WEP security problems is to switch to WPA2. WPA was an intermediate solution for hardware that could not support WPA2. Both WPA and WPA2 are much more secure than WEP. To add support for WPA or WPA2, some old Wi-Fi access points might need to be replaced or have their firmware upgraded. WPA was designed as an interim software-implementable solution for WEP that could forestall immediate deployment of new hardware. However, TKIP (the basis of WPA) has reached the end of its designed lifetime, has been partially broken, and had been officially deprecated with the release of the 802.11-2012 standard.

## Implemented Non-standard Fixes

## WEP2

This stopgap enhancement to WEP was present in some of the early 802.11i drafts. It was implementable on some (not all) hardware not able to handle WPA or WPA2, and extended both the IV and the key values to 128 bits. It was hoped to eliminate the duplicate IV deficiency as well as stop brute force key attacks.

After it became clear that the overall WEP algorithm was deficient (and not just the IV and key sizes) and would require even more fixes, both the WEP2 name and original algorithm were dropped. The two extended key lengths remained in what eventually became WPA's TKIP.

## WEPplus

WEPplus, also known as WEP+, is a proprietary enhancement to WEP by Agere Systems (formerly a subsidiary of Lucent Technologies) that enhances WEP security by avoiding "weak IVs". It is only completely effective when WEPplus is used at both ends of the wireless connection. As this cannot easily be enforced, it remains a serious limitation. It also does not necessarily prevent replay attacks, and is ineffective against later statistical attacks that do not rely on weak IVs.

## Dynamic WEP

Dynamic WEP refers to the combination of 802.1x technology and the Extensible Authentication Protocol. Dynamic WEP changes WEP keys dynamically. It is a vendor-specific feature provided by several vendors such as 3Com.

The dynamic change idea made it into 802.11i as part of TKIP, but not for the actual WEP algorithm.

# WPA (Wi-Fi Protected Access)

Wi-Fi Protected Access (WPA), Wi-Fi Protected Access II (WPA2), and Wi-Fi Protected Access 3 (WPA3) are three security protocols and security certification programs developed by the Wi-Fi Alliance to secure wireless computer networks. The Alliance defined these in response to serious weaknesses researchers had found in the previous system, Wired Equivalent Privacy (WEP).

WPA (sometimes referred to as the draft IEEE 802.11i standard) became available in 2003. The Wi-Fi Alliance intended it as an intermediate measure in anticipation of the availability of the more secure and complex WPA2, which became available in 2004 and is common shorthand for the full IEEE 802.11i (or IEEE 802.11i-2004) standard.

In January 2018, Wi-Fi Alliance announced the release of WPA3 with several security improvements over WPA2.

## WPA

The Wi-Fi Alliance intended WPA as an intermediate measure to take the place of WEP pending the availability of the full IEEE 802.11i standard. WPA could be implemented through firmware upgrades on wireless network interface cards designed for WEP that began shipping as far back as 1999. However, since the changes required in the wireless access points (APs) were more extensive than those needed on the network cards, most pre-2003 APs could not be upgraded to support WPA.

The WPA protocol implements much of the IEEE 802.11i standard. Specifically, the Temporal Key Integrity Protocol (TKIP) was adopted for WPA. WEP used a 64-bit or 128-bit encryption key that must be manually entered on wireless access points and devices and does not change. TKIP employs a per-packet key, meaning that it dynamically generates a new 128-bit key for each packet and thus prevents the types of attacks that compromised WEP.

WPA also includes a Message Integrity Check, which is designed to prevent an attacker from altering and resending data packets. This replaces the cyclic redundancy check (CRC) that was used by the WEP standard. CRC's main flaw was that it did not provide a sufficiently strong data integrity guarantee for the packets it handled. Well tested message authentication codes existed to solve these problems, but they required too much computation to be used on old network cards. WPA uses a message integrity check algorithm called TKIP to verify the integrity of the packets. TKIP is much stronger than a CRC, but not as strong as the algorithm used in WPA2. Researchers have since discovered a flaw in WPA that relied on older weaknesses in WEP and the limitations of the message integrity code hash function, named Michael, to retrieve the keystream from short packets to use for re-injection and spoofing.

## WPA2

WPA2 replaced WPA. WPA2, which requires testing and certification by the Wi-Fi Alliance, implements the mandatory elements of IEEE 802.11i. In particular, it includes mandatory support for CCMP, an AES-based encryption mode. Certification began in September, 2004; from March 13, 2006, WPA2 certification is mandatory for all new devices to bear the Wi-Fi trademark.

## WPA3

In January 2018, the Wi-Fi Alliance announced WPA3 as a replacement to WPA2.

The new standard uses an equivalent 192-bit cryptographic strength in WPA3-Enterprise mode (AES-256 in GCM mode with SHA-384 as HMAC), and still mandates the use of CCMP-128 (AES-128 in CCM mode) as the minimum encryption algorithm in WPA3-Personal mode.

The WPA3 standard also replaces the Pre-Shared Key exchange with Simultaneous Authentication of Equals as defined in IEEE 802.11-2016 resulting in a more secure initial key exchange in personal mode and forward secrecy. The Wi-Fi Alliance also claims that WPA3 will mitigate security issues posed by weak passwords and simplify the process of setting up devices with no display interface.

Protection of management frames as specified in the IEEE 802.11w amendment is also enforced by the WPA3 specifications.

## Hardware Support

WPA has been designed specifically to work with wireless hardware produced prior to the introduction of WPA protocol, which provides inadequate security through WEP. Some of these devices support WPA only after applying firmware upgrades, which are not available for some legacy devices.

Wi-Fi devices certified since 2006 support both the WPA and WPA2 security protocols. WPA2 may not work with some older network cards.

## WPA Terminology

Different WPA versions and protection mechanisms can be distinguished based on the target end-user (according to the method of authentication key distribution), and the encryption protocol used.

### Target Users (Authentication Key Distribution)

### WPA-PERSONAL

Also referred to as WPA-PSK (pre-shared key) mode, this is designed for home and small office networks and doesn't require an authentication server. Each wireless network device encrypts the network traffic by deriving its 128-bit encryption key from a 256-bit shared key. This key may be entered either as a string of 64 hexadecimal digits, or as a passphrase of 8 to 63 printable ASCII characters. If ASCII characters are used, the 256-bit key is calculated by applying the PBKDF2 key derivation function to the passphrase, using the SSID as the salt and 4096 iterations of HMAC-SHA1. WPA-Personal mode is available with both WPA and WPA2.

### WPA-Enterprise

Also referred to as WPA-802.1X mode, and sometimes just WPA (as opposed to WPA-PSK), this is designed for enterprise networks and requires a RADIUS authentication server. This requires a more complicated setup, but provides additional security (e.g. protection against dictionary attacks on short passwords). Various kinds of the Extensible Authentication Protocol (EAP) are used for authentication. WPA-Enterprise mode is available with both WPA and WPA2.

## Wi-Fi Protected Setup

This is an alternative authentication key distribution method intended to simplify and strengthen the process, but which, as widely implemented, creates a major security hole via WPS PIN recovery.

## Encryption Protocol

## Temporal Key Integrity Protocol

The RC4 stream cipher is used with a 128-bit per-packet key, meaning that it dynamically generates a new key for each packet. This is used by WPA.

## CTR mode with CBC-MAC Protocol

The protocol used by WPA2, based on the Advanced Encryption Standard (AES) cipher along with strong message authenticity and integrity checking is significantly stronger in protection for both privacy and integrity than the RC4-based TKIP that is used by WPA. Among informal names are "AES" and "AES-CCMP". According to the 802.11n specification, this encryption protocol must be used to achieve fast 802.11n high bitrate schemes, though not all implementations enforce this. Otherwise, the data rate will not exceed 54 Mbit/s.

## EAP Extensions under WPA and WPA2 Enterprise

Originally, only EAP-TLS (Extensible Authentication Protocol Transport Layer Security) was certified by the Wi-Fi alliance. In April 2010, the Wi-Fi Alliance announced the inclusion of additional EAP types to its WPA- and WPA2- Enterprise certification programs. This was to ensure that WPA-Enterprise certified products can interoperate with one another.

As of 2010 the certification program includes the following EAP types:

- EAP-TLS (previously tested).
- EAP-TTLS/MSCHAPv2.
- PEAPv0/EAP-MSCHAPv2.
- PEAPv1/EAP-GTC.
- PEAP-TLS.
- EAP-SIM.
- EAP-AKA.
- EAP-FAST.

802.1X clients and servers developed by specific firms may support other EAP types. This certification is an attempt for popular EAP types to interoperate; their failure to do so as of 2013 is one of the major issues preventing rollout of 802.1X on heterogeneous networks.

Commercial 802.1X servers include Microsoft Internet Authentication Service and Juniper Networks Steelbelted RADIUS as well as Aradial Radius server. FreeRADIUS is an open source 802.1X server.

## Security Issues

### Weak Password

Pre-shared key WPA and WPA2 remain vulnerable to password cracking attacks if users rely on a weak password or passphrase. WPA passphrase hashes are seeded from the SSID name and its length; rainbow tables exist for the top 1,000 network SSIDs and a multitude of common passwords, requiring only a quick look up to speed up cracking WPA-PSK.

Brute forcing of simple passwords can be attempted using the Aircrack Suite starting from the four-way authentication handshake exchanged during association or periodic re-authentication.

WPA3 replaces cryptographic protocols susceptible to off-line analysis with protocols that require interaction with the infrastructure for each guessed password, supposedly placing temporal limits on the number of guesses. However, design flaws in WPA3 enables attackers to plausibly launch brute-force attacks.

### Lack of Forward Secrecy

WPA and WPA2 don't provide forward secrecy, meaning that once an adverse person discovers the pre-shared key, they can potentially decrypt all packets encrypted using that PSK transmitted in the future and even past, which could be passively and silently collected by the attacker. This also means an attacker can silently capture and decrypt others' packets if a WPA-protected access point is provided free of charge at a public place, because its password is usually shared to anyone in that place. In other words, WPA only protects from attackers who don't have access to the password. Because of that, it's safer to use Transport Layer Security (TLS) or similar on top of that for the transfer of any sensitive data. However starting from WPA3, this issue has been addressed.

### WPA Packet Spoofing and Decryption

Mathy Vanhoef and Frank Piessens significantly improved upon the WPA-TKIP attacks of Erik Tews and Martin Beck. They demonstrated how to inject an arbitrary number of packets, with each packet containing at most 112 bytes of payload. This was demonstrated by implementing a port scanner, which can be executed against any client using

WPA-TKIP. Additionally they showed how to decrypt arbitrary packets sent to a client. They mentioned this can be used to hijack a TCP connection, allowing an attacker to inject malicious JavaScript when the victim visits a website. In contrast, the Beck-Tews attack could only decrypt short packets with mostly known content, such as ARP messages, and only allowed injection of 3 to 7 packets of at most 28 bytes. The Beck-Tews attack also requires Quality of Service (as defined in 802.11e) to be enabled, while the Vanhoef-Piessens attack does not. Neither attack leads to recovery of the shared session key between the client and Access Point.

Halvorsen and others show how to modify the Beck-Tews attack to allow injection of 3 to 7 packets having a size of at most 596 bytes. The downside is that their attack requires substantially more time to execute: approximately 18 minutes and 25 seconds. In other work Vanhoef and Piessens showed that, when WPA is used to encrypt broadcast packets, their original attack can also be executed. This is an important extension, as substantially more networks use WPA to protect broadcast packets, than to protect unicast packets. The execution time of this attack is on average around 7 minutes, compared to the 14 minutes of the original Vanhoef-Piessens and Beck-Tews attack.

The vulnerabilities of TKIP are significant in that WPA-TKIP had been held to be an extremely safe combination; indeed, WPA-TKIP is still a configuration option upon a wide variety of wireless routing devices provided by many hardware vendors. A survey in 2013 showed that 71% still allow usage of TKIP, and 19% exclusively support TKIP.

## WPS PIN Recovery

A more serious security flaw was revealed in December 2011 by Stefan Viehböck that affects wireless routers with the Wi-Fi Protected Setup (WPS) feature, regardless of which encryption method they use. Most recent models have this feature and enable it by default. Many consumer Wi-Fi device manufacturers had taken steps to eliminate the potential of weak passphrase choices by promoting alternative methods of automatically generating and distributing strong keys when users add a new wireless adapter or appliance to a network. These methods include pushing buttons on the devices or entering an 8-digit PIN.

The Wi-Fi Alliance standardized these methods as Wi-Fi Protected Setup; however the PIN feature as widely implemented introduced a major new security flaw. The flaw allows a remote attacker to recover the WPS PIN and, with it, the router's WPA/WPA2 password in a few hours. Users have been urged to turn off the WPS feature, although this may not be possible on some router models. Also, the PIN is written on a label on most Wi-Fi routers with WPS, and cannot be changed if compromised.

WPA3 introduces a new alternative for configuration of devices that lack sufficient user interface capabilities by allowing nearby devices to serve as an adequate UI for network provisioning purposes, thus mitigating the need for WPS.

## MS-CHAPv2 and lack of AAA Server CN Validation

Several weaknesses have been found in MS-CHAPv2, some of which severely reduce the complexity of brute-force attacks making them feasible with modern hardware. In 2012 the complexity of breaking MS-CHAPv2 was reduced to that of breaking a single DES key, work by Moxie Marlinspike and Marsh Ray. Moxie advised: "Enterprises who are depending on the mutual authentication properties of MS-CHAPv2 for connection to their WPA2 Radius servers should immediately start migrating to something else".

Tunneled EAP methods using TTLS or PEAP which encrypt the MSCHAPv2 exchange are widely deployed to protect against exploitation of this vulnerability. However, prevalent WPA2 client implementations during the early 2000s were prone to misconfiguration by end users, or in some cases (e.g. Android), lacked any user-accessible way to properly configure validation of AAA server certificate CNs. This extended the relevance of the original weakness in MSCHAPv2 within MiTM attack scenarios. Under stricter WPA2 compliance tests announce alongside WPA3, certified client software will be required to conform to certain behaviors surrounding AAA certificate validation.

## Hole196

Hole196 is vulnerability in the WPA2 protocol that abuses the shared Group Temporal Key (GTK). It can be used to conduct man-in-the-middle and denial-of-service attacks. However, it assumes that the attacker is already authenticated against Access Point and thus in possession of the GTK.

## Predictable Group Temporal Key

In 2016 it was shown that the WPA and WPA2 standards contain an insecure expository random number generator (RNG). Researchers showed that, if vendors implement the proposed RNG, an attacker is able to predict the group key (GTK) that is supposed to be randomly generated by the access point (AP). Additionally, they showed that possession of the GTK enables the attacker to inject any traffic into the network, and allowed the attacker to decrypt all internet traffic transmitted over the wireless network. They demonstrated their attack against an Asus RT-AC51U router that uses the MediaTek out-of-tree drivers, which generate the GTK themselves, and showed the GTK can be recovered within two minutes or less. Similarly, they demonstrated the keys generated by Broadcom access daemons running on VxWorks 5 and later can be recovered in four minutes or less, which affects, for example, certain versions of Linksys WRT54G and certain Apple AirPort Extreme models. Vendors can defend against this attack by using a secure RNG. By doing so, Hostapd running on Linux kernels is not vulnerable against this attack and thus routers running typical OpenWrt or LEDE installations do not exhibit this issue.

## KRACK Attack

In October 2017, details of the KRACK (Key Reinstallation Attack) attack on WPA2 were published. The KRACK attack is believed to affect all variants of WPA and WPA2; however, the security implications vary between implementations, depending upon how individual developers interpreted a poorly specified part of the standard. Software patches can resolve the vulnerability but are not available for all devices.

## Dragonblood Attack

In April 2019, serious design flaws in WPA3 were found which allow attackers to perform downgrade attacks and side-channel attacks, enabling brute-forcing the passphrase, as well as launching denial-of-service attacks on Wi-Fi base stations.

## Security Issues to Wireless Networks

Network security issues, whether wired or wireless, fall into three main categories: availability, confidentiality and integrity:

- Confidentiality: Is the information being sent across the network transmitted in such a way that only the intended recipient(s) can read it.

- Integrity: Is the information reaching the recipient intact.

- Availability: Is the network available to users whenever it is supposed to be.

## Confidentiality

The main way to ensure that data is not disclosed to unauthorised users is by encrypting it during transit, and wireless networks are able to do this in just the same way as wired networks. However, encryption is meaningless without authentication, since an unauthorised user could authenticate themselves onto the network and then be given the key with which to decrypt the data.

The traditional model for authorisation is to have some form of centralised system which stores access control lists. This model is fine for use in networks which have a relatively static set of users, and so is suitable for Wi-Fi, but in other networks such as Bluetooth networks, which are much more ad-hoc in nature, this approach becomes impractical. In ad-hoc networks, not only does the dynamically changing set of users make updating access control lists infeasible in terms of cost, but there is also no guarantee that these devices would be able to access any central system. In these systems, a better approach is to form secure transient associations between devices, where the decision on who to trust is made either by each device, or by one master device which instructs the slave devices on how to behave.

There is a lot of interest in this model for applications such as controlling police weapons. In America, a large proportion of injuries to police officers come from stolen police guns. If each officer had a very short range ring (the master) associated with the gun (the slave) it would prevent anyone other than that officer from using the weapon.

## Integrity

Because packets of data in wireless networks are sent through the air, they can be intercepted and modified quite easily by malicious users. This means that wireless networks are more vulnerable to attacks on the integrity of data. However, the current methods used by wired networks to ensure the integrity of packets, such as checksums, are perfectly adequate for ensuring the integrity of packets in wireless networks, and so no novel solutions have been adopted.

## Availability

Wireless networks are particularly susceptible to DoS (Denial of Service) attacks. Unlike wired networks, which require the attacker to be physically connected to the network in some way before they can launch such an attack, with wireless networks an attacker only has to be within a certain range of the network (usually 100m) to be able to launch such an attack. These kinds of attacks are particularly difficult to stop since network providers want to allow legitimate users to initiate communications with the network, and cannot stop malicious users from exploiting this to cause a denial of service.

Another way in which malicious users can potentially restrict the availability of the wireless networks is through radio jamming. This involves sending out a lot of noise on the same frequency as the network uses. However, there are techniques, such as frequency hopping which can make this kind of attack more difficult. Also, this threat is less relevant in the non-military world since the 'jammer' could be reported to the police and arrested.

One kind of attack on the availability of wireless networks which has arisen in the last few years is battery exhaustion attacks. Because many wireless network devices are portable and therefore battery powered, malicious users can repeatedly send messages to the device. This prevents it from going into its sleep mode, and the battery runs down much faster.

## References

- Wireless-Networking-Security: cs.bham.ac.uk, Retrieved 23 May, 2020

- Harwood, Mike (29 June 2009). "Securing Wireless Networks". CompTIA Network+ N10-004 Exam Prep. Pearson IT Certification. p. 287. ISBN 978-0-7897-3795-3. Retrieved 9 July 2020

- Mobile-computing-security-issues, mobile-computing: tutorialspoint.com, Retrieved 4 July, 2020

- Vanhoef, Mathy; Piessens, Frank (May 2013). "Practical Verification of WPA-TKIP Vulnerabilities" (PDF). Proceedings of the 8th ACM SIGSAC symposium on Information, computer and communications security. ASIA CCS '13: 427–436. doi:10.1145/2484313.2484368

# PERMISSIONS

All chapters in this book are published with permission under the Creative Commons Attribution Share Alike License or equivalent. Every chapter published in this book has been scrutinized by our experts. Their significance has been extensively debated. The topics covered herein carry significant information for a comprehensive understanding. They may even be implemented as practical applications or may be referred to as a beginning point for further studies.

We would like to thank the editorial team for lending their expertise to make the book truly unique. They have played a crucial role in the development of this book. Without their invaluable contributions this book wouldn't have been possible. They have made vital efforts to compile up to date information on the varied aspects of this subject to make this book a valuable addition to the collection of many professionals and students.

This book was conceptualized with the vision of imparting up-to-date and integrated information in this field. To ensure the same, a matchless editorial board was set up. Every individual on the board went through rigorous rounds of assessment to prove their worth. After which they invested a large part of their time researching and compiling the most relevant data for our readers.

The editorial board has been involved in producing this book since its inception. They have spent rigorous hours researching and exploring the diverse topics which have resulted in the successful publishing of this book. They have passed on their knowledge of decades through this book. To expedite this challenging task, the publisher supported the team at every step. A small team of assistant editors was also appointed to further simplify the editing procedure and attain best results for the readers.

Apart from the editorial board, the designing team has also invested a significant amount of their time in understanding the subject and creating the most relevant covers. They scrutinized every image to scout for the most suitable representation of the subject and create an appropriate cover for the book.

The publishing team has been an ardent support to the editorial, designing and production team. Their endless efforts to recruit the best for this project, has resulted in the accomplishment of this book. They are a veteran in the field of academics and their pool of knowledge is as vast as their experience in printing. Their expertise and guidance has proved useful at every step. Their uncompromising quality standards have made this book an exceptional effort. Their encouragement from time to time has been an inspiration for everyone.

The publisher and the editorial board hope that this book will prove to be a valuable piece of knowledge for students, practitioners and scholars across the globe.

# INDEX